ANIMAL REPRODUCTION

BY THE SAME AUTHOR

Animal Weapons

ANIMAL REPRODUCTION

Philip Street

Drawings By
Nancy Lou Gahan

DAVID & CHARLES
NEWTON ABBOT LONDON

0 7153 6376 X

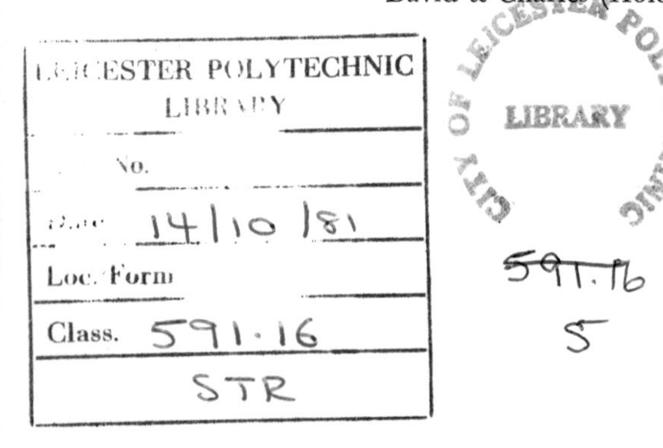

Set in 11 on 13-point Baskerville
and printed in Great Britain
by W J Holman Limited Dawlish
for David & Charles (Holdings) Limited
South Devon House Newton Abbot Devon

CONTENTS

LIST OF ILLUSTRATIONS

CHAPTER ONE

INTRODUCING THE VERTEBRATES

Modern classification recognises about twenty main groups of animals called phyla, but many of these are obscure groups containing comparatively few animals, whose very existence is often known only to the expert. Among them, however, only two have solved completely the problems of producing animals that are fully adapted to a terrestrial life; and these have been evolved from fully aquatic ancestors, for life originated in the waters and was for long confined to them. The phylum Arthropoda includes the vast assemblage of insects and spiders; the phylum Chordata contains the reptiles, birds and mammals, and these two groups between them dominate life on land in every part of the world.

The first vertebrates were undoubtedly fish, but in order to understand their structure and how they were evolved it is necessary to look back to the more primitive members of the phylum Chordata. These existed before the first vertebrate appeared, and it was from one of their number that it developed. The feature that links all the chordates together into one phylum is the possession of a notochord, a stiffened rod of cells running down the length of the back and providing the main support for the body. In the more primitive chordates this notochord is the only skeletal support, but in the vertebrates it has become reinforced by the development around it of cartilage or bone in the form of separate cylinders called vertebrae, the complete series being the vertebral column from which they derive their name. In the higher vertebrates these vertebrae have become so well developed in the adult that they have entirely replaced the softer notochord, but this still develops in the embryo and is only subsequently obliterated.

The ancestral chordate from which the first fish was derived between three and four hundred million years ago must, therefore, have possessed a notochord, and probably had the general shape of a fish. Fortunately the few remaining primitive chordates do between them give us a considerable amount of information enabling us to piece together the probable series of events that gave rise to the vertebrates.

The phylum Chordata is divided into four subphyla, three consisting of primitive chordates and the fourth comprising the vertebrates. One of the subphyla, the Cephalochordata, contains the little marine creature called *Amphioxus,* which is believed to have altered little for three hundred million years or more and is thought to be similar in structure to the early chordates from which the first fish were derived. *Amphioxus* thus shows us the basic plan from which the future structure of the vertebrates was derived.

Despite its age and its comparatively primitive structure *Amphioxus* must nevertheless be regarded as a successful type, for it is widely distributed and found in almost every part of the temperate and tropical world. It lives in the shallow offshore waters, where it spends almost the whole of its time buried just below the surface of the sand for protection. It is a small creature, somewhat less than two inches long and of general fish-like shape. The notochord is prominent running along the length of its back, with a hollow nerve chord corresponding to the vetrebrate spinal chord lying immediately above it. Along each side of the body there is a series of powerful muscles continuing right to the tail end. It is by the alternate contraction and relaxation of these that *Amphioxus* is able to swim with a wriggling motion similar to that of a fish.

There is a mouth but no jaws, and the creature feeds by filtering out from the sea water minute organisms suspended in it. A continuous current of water is drawn into the mouth by the rhythmical beating of countless hair-like cilia surrounding it. This water current passes to the back of the throat, or pharynx, whose side walls are pierced by hundreds of minute slits, usually referred to as gill slits. The water passes through these slits, which act as filters and strain off the food particles, the filtered water passing on to escape to the exterior while the trapped food is swallowed. The gill slits do not open separately along the sides of the animal, but they all lead into a special cavity which opens to the exterior by a single pore. Gills in most aquatic animals are respiratory organs whose purpose is to extract oxygen from the water. The principal function of the so-called gills of the primitive chordates, however, is to collect suspended food particles, and any respiratory function they may perform is purely incidental.

While many of the features of *Amphioxus* are reminiscent of those of the fishes, there is little about the adult structure of the sea-squirts, which belong to the subphylum Urochordata, to suggest even the remotest relationship with the vertebrates. The body of a typical sea-squirt consists of a U-shaped tube, one arm of which is much wider than the other. This wider arm is the front end of the animal, and consists almost entirely of a pharynx perforated with a very large number of tiny gill slits similar to those of *Amphioxus*. As with *Amphioxus*, too, their function is to extract minute food organisms from a continuous current of water that is drawn in through the mouth. This is accomplished by the beating of cilia lining the gill slits themselves. Surrounding the whole body of the animal is a non-living tunic composed of a carbohydrate called tunicin, which is closely related chemically to cellulose. The tunic has two openings corresponding to the front and hind ends of the body of the animal. The possession of a tunic has given the Urochordata their alternative name, tunicates.

Apart from the pharynx with its many gill slits, the adult sea-squirts have no other features to betray their chordate affinities. The indisputable evidence that they do indeed belong to this phylum is found in their development. The egg of a sea-squirt hatches to a free-living larva which is of tadpole shape and able to swim by the side-to-side movement of a long muscular tail. This ascidian tadpole, as it is called, has a well developed notochord running the whole length of the body and the tail and a typical hollow chordate nerve chord above it.

The ascidian tadpole is not only very like *Amphioxus* or an amphibian tadpole in fundamental structure, but its development from the egg follows very closely the method of development of *Amphioxus* and the frog. This is, however, only a transitory larval stage, which after a few days of free life near the surface of the sea settles on the sea bed, undergoes a metamorphosis and grows into an adult sea-squirt. The purpose of this ascidian tadpole larva is to distribute the offspring of a creature which in the adult stage cannot move about and so could not colonise new territory.

Supposing that *Amphioxus* was evolved from an ancestor similar to the tunicates, it must have inherited the structure of the larva and not of the adult. The tunicates, fortunately, do show us how this stage of evolution might have occurred. In one group of the Urochordata, the Larvacea or Appendicularia, the tadpole that hatches from the egg never undergoes a metamorphosis: instead it retains its fish-like structure throughout life, becoming sexually mature and able to breed. What has happened is that the relative rates of development of the reproductive organs and the rest of the body have become altered in such a way that sexual maturity is attained before the body is ready to metamorphose, and once maturity has been achieved metamorphosis is suppressed.

The condition in which a larva becomes sexually mature by a speeding up of the rate of development of the reproductive organs relative to the rate of development of the rest of the body is known as neoteny. As we shall see later, neoteny occurs in the Amphibia—the axolotl is a salamander tadpole that becomes sexually mature and so never undergoes metamorphosis, remaining in the tadpole stage for its whole life.

Although we have no definite proof of events that may have taken place as long as four hundred million years ago, it seems fairly certain that the earliest chordate was probably a seden-

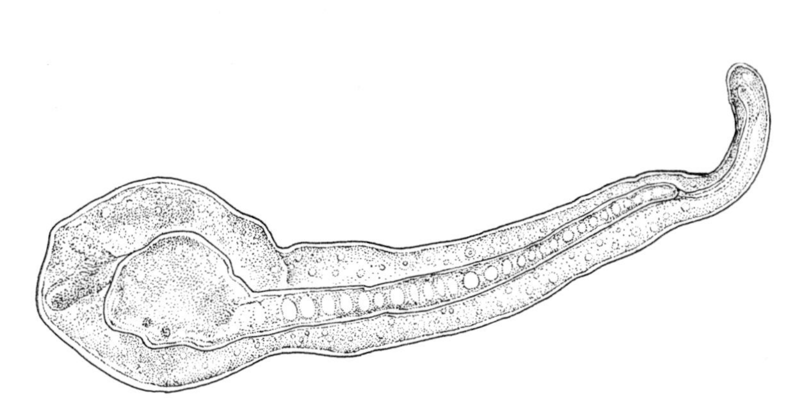

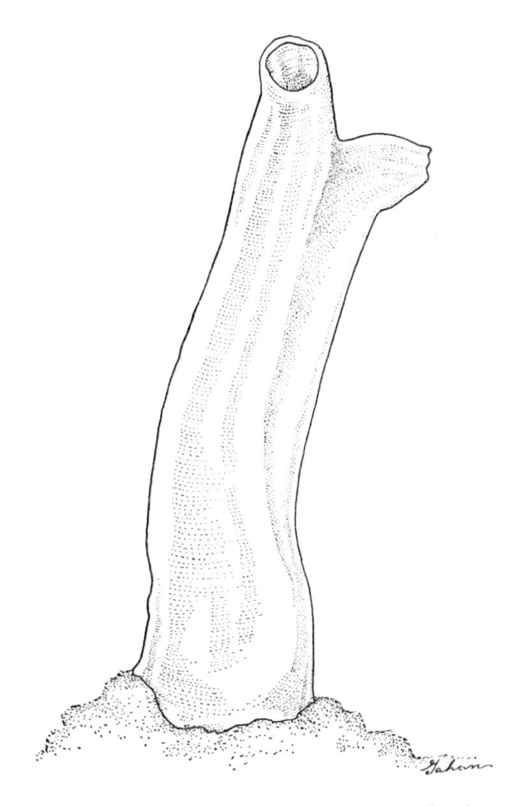

Ascidian tadpole of the Urochordata (sea-squirt tadpole) and a full-grown sea-squirt (not to scale to each other)

tary organism feeding by means of a perforated pharynx and including a free-living tadpole stage in its life history to provide the possibility of distribution. At this tadpole stage it was an active creature, which had developed a relatively powerful musculature to enable it to swim, a notochord to provide support for its body and a central nerve chord above it. By a process of neoteny this tadpole larval stage became the adult stage, and the sedentary phase disappeared from the life cycle. This produced a creature similar to *Amphioxus* and showing all the fundamental chordate features. From an *Amphioxus*-like ancestor it is not difficult to deduce the probable line of evolution of the first vertebrate.

Having considered the likely structure of the earliest chordates from which the whole phylum was subsequently developed, we must now turn to the equally interesting and important question of the origin of the chordates themselves, to see whether there is any evidence to establish from which invertebrate group they were evolved at an even earlier date. A valuable clue enabling us to give the probable answer to this question comes from a third chordate subphylum, the Hemichordata. Two kinds of animals are included in this group. One kind are worm-like, burrowing creatures, while the other kind are sedentary. Details of their structure are unimportant, and all we need to know is that both kinds contain sufficient evidence to show that they are chordates, possessing an undoubted notochord, though they are much more primitive than the members of the Urochordata and the Cephalochordata. Their importance lies in their development.

From the egg of the Hemichordata a tiny ciliated tornaria larva develops which is virtually indistinguishable from the pluteus larvae produced from the eggs of some members of the invertebrate phylum Echinodermata, the phylum that comprises the starfishes, sea-urchins, sea-cucumbers and feather stars. In zoology it is a well-established principle that the development of an animal is often a much more reliable guide to its relationships than its adult structure, and virtual identity of structure between the tornaria larva of the Hemi-

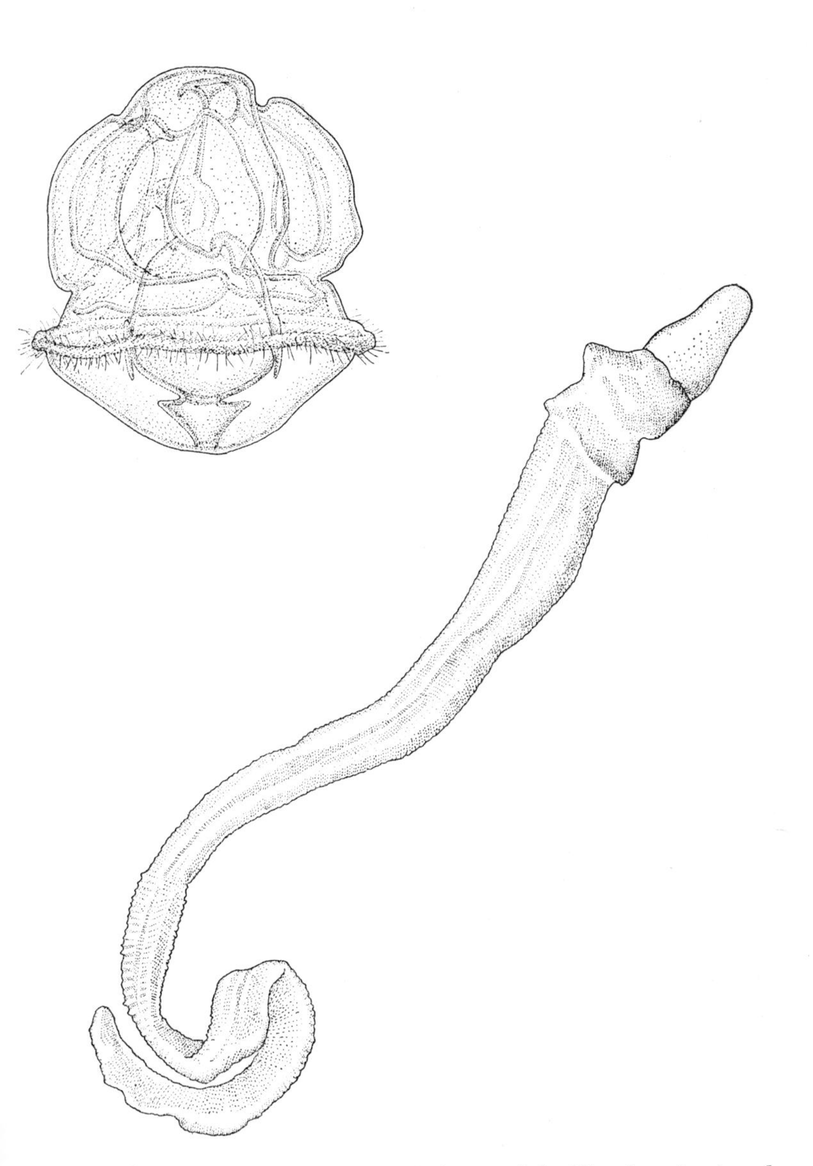

Acorn worm larva (ciliated tornaria larva of the Hemicordata) and full-grown acorn worm (not to scale to each other)

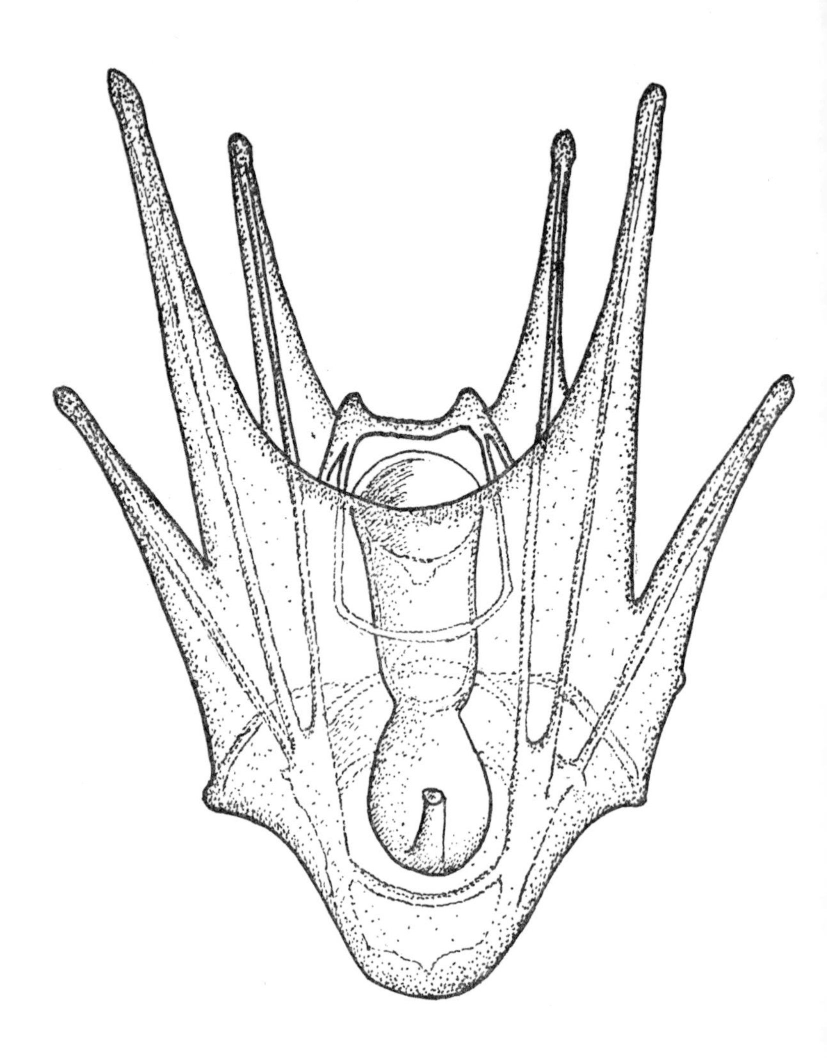

Echinopluteus larva of the sea-urchin

chordata and the pluteus larvae typical of the echinoderms can be taken as reliable evidence of the close relationship of the two groups. Of course the primitive echinoderms from which the first chordate arose would have been very different from the modern representatives of the phylum. After all they, too, have had at least four hundred million years to evolve, and there is no reason to suppose that they are not today drastically different from their remote and primitive ancestors.

We are now in a position to hazard a guess as to the events that gave rise to the early evolution of the chordates. The early echinoderm ancestor evolved a free-swimming tadpole-like stage succeeding the pluteus larval stage which hatched from the egg, and the tadpole in turn gave rise by metamorphosis to a sedentary adult stage in which filter feeding by means of pharyngeal gills was developed. Subsequently this method of feeding was extended back to the tadpole stage. At a later period the pluteus larva became suppressed and the egg then developed directly into the tadpole, which because it moved about actively had developed the musculature and the notochord that were to become fundamental characteristics of the phylum. Finally, as we have already seen, neoteny resulted in the loss of the sedentary adult stage and its replacement by the now mature fish-like tadpole stage.

The evolution of these early chordates was the essential preliminary to the development of the vertebrates. From both the zoological and human points of view the subsequent evolution of the fish, amphibians and reptiles was of supreme importance, because it was from the reptiles that the birds and the mammals, including man, were eventually evolved.

The first undoubted vertebrates were thus similar in essential structure to *Amphioxus*, but showed additional features that stamped them as vertebrates. Most important of these was the strengthening of the notochord by the development around it of cartilage, which in later forms became organised as definite vertebrae and later still became made of bone instead of cartilage. The nerve chord remained above the notochord, but at its front end it became enlarged to form the brain, which was enclosed and protected by a special development of the anterior vertebrae to form the skull.

In the earliest vertebrates food was probably still collected by filtering water through numerous gills in the pharynx, but this method was soon abandoned. The gills, however, were retained and took over the responsibility of supplying the animals with the oxygen they needed. The strengthening bars between a few of the gills at the front end of the pharynx were drastically altered to form the skeleton of the jaws, which developed as the new way of obtaining food.

These advances did not of course all occur at once. It was in fact only after many millions of years of steady change that all of them had become incorporated. The first vertebrates to appear in the fossil records were fish without jaws, and a few of these remain with us today to form the most primitive of all the living vertebrates, comprising the class Cyclostomata.

We are now in a position to summarise the classification of the living chordates. Until recently all fishes except the primitive jawless lampreys and hag-fishes were included in a single class Pisces, but today it is recognised that this class contained three different kinds of fishes which were so different and had been separate groups for such a long time that they should be assigned to three separate classes.

PHYLUM CHORDATA

Subphylum Hemichordata
Subphylum Cephalochordata
Subphylum Urochordata
Subphylum Vertebrata

Class 1 Cyclostomata: lampreys and hag-fishes
2 Elasmobranchii: sharks, skates and rays
3 Actinopterygii: bony fishes
4 Choanichthyes: lung-fishes and coelacanths
5 Amphibia: frogs, toads, newts and salamanders
6 Reptilia: snakes, lizards, crocodiles and tortoises
7 Aves: birds
8 Mammalia: mammals

CHAPTER TWO

FISHES

The lampreys and the hag-fishes, which comprise the small class of jawless fishes or Cyclostomata, differ considerably in their reproductive methods.

The life history of the lampreys is particularly interesting. Adult sea lampreys and river lampreys spend most of their time in the sea but migrate into rivers to breed. When they are approaching maturity, in the late summer or autumn, they swim up the rivers, after which they never feed again. During the winter their reproductive organs gradually ripen in preparation for breeding the following spring. Spawning is preceded by a primitive form of nest-building. The mature lampreys assemble in considerable numbers on stretches of the river where the water is shallow and fast moving and the bottom is stony. Nests, which are merely shallow depressions in the river bed, are made by removing stones one at a time with the suckers. This is usually done by the males. The animals then come together in pairs, and the eggs and spermatozoa are shed simultaneously while the members of a pair are intertwined.

Development of the lamprey egg is very similar to that of the egg of a frog, and from it hatches an ammocoete larva, a tiny transparent creature less than half an inch in length. This larva differs in several important respects from the adult lampreys, and is more like an adult *Amphioxus*. It has numerous gills lining a well-developed pharynx and feeds by filtering minute organisms from a continuous current of water, which is kept moving by muscular action.

For most of its time the ammocoete larva remains buried in

the mud on the bed of the river and is thus seldom seen. There is no sucker around the mouth as there is in the adult, and the eyes remain beneath the skin and are consequently functionless. Lampreys remain in this larval state for several years before finally undergoing metamorphosis, which converts them to adults. By this time they will be about six inches long. They now forsake the rivers for the sea, where they will remain until it is time for them in their turn to ascend the rivers to breed. Once a lamprey has spawned it dies, so that each individual has only one breeding season during the course of its life.

In contrast to the sea and river lampreys there is a third group known as brook lampreys, which are smaller than the other two and never migrate down to the sea, but spend the whole of their lives in fresh water. Like the other species they spend most of their time in the larval ammocoete stage, lying buried in the mud on the river bed, so that although they are quite common in many rivers and streams they are seldom seen. After about three years they at last undergo metamorphosis, usually in the late summer and autumn. Once they have become adults they do not feed, but like the migratory lampreys live through the winter while their reproductive organs ripen in preparation for the spring spawning, after which they die.

Unlike the lampreys the hag-fishes are entirely marine. The females lay small numbers of large eggs well supplied with yolk and enclosed in horny shells. There is no larval stage, the young which hatch from them having similar structure to the adults.

Apart from the Cyclostomata, all the remaining vertebrates possess jaws; they comprise the super-class Gnathostomata, which contains seven classes. Three of these are fishes, the first being the class Elasmobranchii or Chondrichthyes, the cartilaginous fishes. Within this class there are three main groups, the skates and rays, the dog-fishes and sharks, and the curious Chimaera and its relatives.

Despite the fact that the elasmobranchs are considered to be more primitive than the bony fishes, their method of reproduction is more advanced than that of most of the higher fishes. The majority of these, as we shall see, spawn into the sea, so that their eggs and young are completely at the mercy of the higher vertebrates, and if they are to be laid they are method entails, each female has to produce an immense number of eggs to ensure that a few young survive to maturity.

The elasmobranchs have gone a long way towards solving this problem. Their eggs are fertilised internally, like the eggs of the higher vertebrates, and if they are to be laid they are covered with a protective shell after having been provided with a large store of yolk. Alternatively they are retained within the oviduct where they develop until they are ready to emerge as young fish. These eggs are also provided with a store of yolk.

About half the total number of skates and rays belong to a single family, the Rajidae, and these are the ones commonly known as skates. Their eggs are particularly large, and after fertilisation each is covered with a roughly rectangular egg case which has a short horn projecting from each corner. When ready to be laid the egg cases are not shed indiscriminately into the water, but are buried in the sandy bed of the sea in fairly deep offshore waters, where they can develop in comparative safety. Skates are believed to lay their eggs two at a time, one from each oviduct, with an interval of several hours between the laying of successive pairs while the egg capsules are being formed round the eggs. The horns are hollow, and allow water to enter and leave the egg case and so provide for the respiratory needs of the developing fish. The fish itself develops on the top of the large egg, the remainder forming an enormous yolk sac suspended from its abdomen, which gradually diminishes as its stored food material is utilised by the developing embryo. Full development takes several months, after which the young fish emerges from the egg case to lead an independent life. The empty cases, washed up on the shore, are commonly known as mermaids' purses.

In contrast the rays, which produce similar large yolky eggs, retain these within their oviducts until they are fully developed. As development proceeds the food stored in the yolk is supplemented by a nutritious milky fluid secreted by special projections from the walls of the oviducts. This fluid is absorbed through the mouth and gill filaments of the developing embryo.

As with the skates, the embryo develops at the top of the egg, so that in the early stages a tiny developing fish has attached to its abdominal region a relatively enormous yolk sac. As development continues, so the relative sizes of the embryo and the yolk sac change. By the time the young hatches or is born the yolk sac is almost completely absorbed.

Most sharks retain their eggs within the oviduct while they develop. Dog-fishes, the Port Jackson shark and the largest of all the sharks, the whale shark, however, lay eggs contained in egg cases. These are relatively more elongated than those produced by the skates, and at each of the four corners there is a long coiling hollow tendril instead of the short horn of the skate egg case. The eggs are not buried in the sand, but are attached to seaweeds by the tendrils. The whale shark, in keeping with its size, produces an egg case about a foot in length.

Most sharks that retain their eggs within the oviduct during development produce large litters, the tiger shark for example giving birth eventually to as many as sixty pups, each about two feet long. Generally the developing embryos use up their store of yolk and are then provided with additional nutritive fluid similar to that provided by the rays for their embryos and produced from special filaments projecting from the oviduct walls.

Some species, however, produce only two embryos at one time, and these are relatively much larger at birth than those of species that produce large numbers. Reproduction in the common sand shark, *Odontaspis taurus*, is particularly unusual. Although only one of the two ovaries is functional, a condition quite common among the larger species of sharks, it produces initially two batches of about twenty eggs, each batch being enclosed in a capsule. These come to rest one in each of the two oviducts. But of the twenty or so eggs in each capsule only one develops into an embryo, the remainder providing a source of food for it. The eggs of the sand shark are only about the size of a pea, which is exceptionally small for a shark's egg.

When the two embryos have used up their initial supply of eggs they break out of their capsules and so come to lie free in the oviducts. The ovary has remained active, and produces a steady supply of mature eggs which pass down the oviduct to continue nourishing the embryos. Throughout development the embryos face inwards, as though waiting to receive this constant supply of eggs coming down to them from the ovary.

Not until they are full grown and ready to be born do they turn round to face the openings of the oviducts to the exterior. By this time they will measure some thirty inches in length, approaching one-third of the adult length, which averages nine feet. Development takes about a year to complete, and when the young sharks are born the stomach of each is crammed with upward of two pounds of egg yolk.

Turning round before they are born probably makes the exit much smoother than a tail first exit—a shark's skin stroked from head to tail is smooth, but stroked in the opposite direction it is very abrasive.

Knowledge of the reproductive habits of many shark species is very sketchy, so although there may well be others, the only species known to have similar habits to the sand shark, feeding its developing embryos on copious supplies of eggs, is the porbeagle, *Lamna nasus*. Often it rears two embryos in each oviduct. This is a smaller species than the sand shark, averaging five feet in length. At birth the pups may be more than twenty inches long, their stomachs bulging with their last meal of eggs.

Certain shark species, notably the blue shark and the hammer-heads, have evolved an even more advanced method of feeding their developing embryos. As soon as the embryos have used up their initial stores of yolk they break out of the egg capsules, and the wall of the abdomen and that of the oviduct come into intimate contact to form a placenta, a structure that reaches its highest development in the mammals. In the placenta blood vessels belonging to the maternal circulation come into intimate contact with blood vessels belonging to the embryo circulation. The walls separating the two sets of blood vessels are so thin that oxygen and food materials can pass readily through them from the maternal to the embryo blood, and carbon dioxide and other waste materials can pass in the reverse direction. In this way the female shark continues to feed its growing embryos until they are fully developed and ready to be born.

Any animal that retains its fertilised eggs within the oviduct while they develop is generally referred to as viviparous, but it is possible to distinguish two types of viviparity: true viviparity, in which there is some kind of placental link between the mother and her developing offspring, and ovoviviparity, where there is no such intimate link. Animals that lay their eggs for external development are described as oviparous. Thus the blue shark is viviparous while the sand shark is ovoviparous. Throughout this book, however, I have used the term viviparous for all examples where there is internal development of the egg, making clear those examples where there is some kind of placenta.

Because, unlike the majority of bony fishes, all elasmobranchs indulge in internal fertilisation, it is not surprising that we find the males provided with copulatory organs. The inner edges of the hind or pelvic fins are rolled up to form tubes commonly known as claspers, so called because it was at one time believed that they were merely used by the males to clasp the females at mating. Normally they are directed backwards, and in mature males they extend beyond the hind limits of the fins themselves. Prior to mating, however, they become stiffened and erect, and are bent forward, the inner ends coming into contact with the external openings of the sperm ducts. In this way the seminal fluid as it is discharged flows into these tubes, the other ends meantime being inserted into the oviducts of the females.

With their reputation for ferocity, it is not surprising that shark courtship is a potentially lethal occupation. Perhaps it is to reduce the risk to the females with which they are to mate that male sharks cease to feed some time before the breeding season, so reducing their vitality. The females, however, do not undergo a similar fast, and as they are on the average larger than the males it may well be that the danger to the males is greater than that to the females.

Probably the potential aggressiveness of sharks towards one another explains certain behaviour patterns which seem designed to keep various categories apart. Before giving birth to their young, viviparous and ovoviparous females migrate from the deeper waters in which they normally live to shallower nursery grounds nearer to the coast. The females do not feed while they are on these nursery grounds, but soon after giving birth to their young they return to deeper waters. The males apparently never come inshore to menace the nursery grounds.

The young sharks, born in spring or early summer, remain on the breeding grounds feeding and growing until the winter, when they may leave for deeper waters as a mixed sex population. Outside the breeding season the adult males and females seem to remain in separate populations, the sexes only coming together for a short time at the breeding season. All these devices for separation probably serve to minimise the effects of sharks' cannabalistic tendencies.

Despite the undoubted success of the elasmobranchs, as shown both by their variety of species and numbers of individuals, it is nevertheless the bony fishes that are the dominant animals in the sea, since they greatly exceed the cartilaginous fishes in numbers of species and in total population. Some 3,000 species of elasmobranchs are known, but more than 20,000 species of bony fishes are recognised. The total populations of some of the more successful species reach almost astronomical proportions. In the North Atlantic alone, for example, the total herring population is believed to be not less than one million million. Whereas, too, the elasmobranchs are confined to the seas, bony fishes have colonised fresh waters as successfully as they have done the seas.

For the origin of the bony fishes, forming the class Actinopterygii, we must go back almost as far as the origin of the elasmobranchs. Not much later than one group of placoderms gave rise to the first elasmobranchs, the members of another placoderm group became modified to produce the ancestral actinopterygians, the bony fishes, and yet a third gave rise to the class Crossopterygii, the lung fishes. Thus the main lines of fish evolution for the next three hundred million years were laid down.

Although there are four orders in the class Actinopterygii, the vast majority of these belong to the most recent order, the Teleostei. The other three orders are the Palaeoniscoidei, comprising the bichirs which inhabit sluggish muddy rivers in various parts of Africa, the Chondrostei, which are the sturgeons and related species, and the Holostei, comprising the freshwater bowfins and gar-pikes which live in North American lakes and rivers. Only two of these types deserve special mention from the point of view of their methods of reproduction.

The bowfin, *Amia calva*, shows surprisingly complex breeding habits, the male playing a more prominent part in the care of the eggs and young than the female. At the approach of spring the fish come together in pairs. The male excavates a shallow nest in the bed of the lake or river, and into this the female deposits her eggs, which are fertilised by the male as they are laid. Being sticky they cling to the bottom of the nest and to any vegetation which may be growing around the edge, and are thus prevented from being carried off by currents or other disturbances in the water.

From now on the developing eggs are the male's responsibility, and he ensures sufficient supplies of oxygen for them by swimming around them constantly, thus keeping the water moving. Within a day or two they have hatched, but some strong instinct compels the fry to stay close together, and the male stays with them. Throughout their first summer the male swims around always accompanied by his swarm of offspring. Not until the autumn, when they have become well grown young fish about four inches in length, do they finally forsake their parent and swim off to begin an independent life, and by this time they are well able to take care of themselves.

The sturgeons, like salmon and sea lampreys, spend much of their life in the sea, returning to the rivers to spawn. The chief interest in their reproduction is that their eggs form the basis of caviar, by any standards one of the most expensive of all human foods. Typical sturgeons are confined to the rivers and seas of Europe and the Near East, occurring in the greatest abundance in certain Russian rivers and in the Danube. On these rivers there is great fishing activity at the time of the spring spawning migrations into the rivers. The largest species is the beluga, *Huso huso,* which may reach a length of up to thirty feet and weigh well over a ton. A female of this weight may well yield over two hundredweight of eggs.

The teleosts, comprising the vast majority of the bony fishes, can conveniently be divided into the marine and freshwater groups. Those that live in the sea include the various kinds of food fishes that are of great economic importance, and it is with these that we shall deal first.

The most important of all the food fishes in the northern hemisphere is the cod, *Gadus morrhua,* which accounts for nearly half the fish landed by the fishing fleets in the northern regions. It belongs to the important group of demersal fish, which live and feed on the sea bed. Like all marine fish, its distribution is determined by the depth and quality of the sea bed, the temperature of the water and the position of its spawning grounds. Outside the spawning season cod can be caught almost anywhere in the northern and temperate seas of Europe, Iceland and Newfoundland, wherever the depth of the water is between ten and one hundred fathoms, though it is most abundant where the water is less than forty fathoms deep. From about April until the early autumn the fish roam at large. There are, however, distinct stocks of cod, and while a few isolated individuals may swim great distances, leaving their own stock to join another one, in general the cod in the various fishing regions belong to separate populations. In the autumn each population begins to concentrate over its chosen spawning grounds, and this goes on until the end of the year, by which time the concentration of fish on these grounds is very high.

Spawning takes place from January until March, and it is at these times that the fishermen make their largest catches. These catches consist almost entirely of large mature fish of from four to five years old and upwards. The smaller immature fish, of course, do not migrate to the breeding grounds. For all the main cod stocks, the breeding grounds are well known and are all similar in quality. The depth of the water is between twenty and thirty fathoms. The North Sea stock assemble mostly in the northern and central areas, and their spawning grounds occur wherever the water is between these depths. Another distinct stock has spawning grounds around the Lofoten Islands off Norway. The Icelandic cod spawn to the south and south-west of the island, whereas during the rest of the year they can be found on the shelf anywhere around the island.

The cod is one of the most prolific of all fish. A single female may shed as many as eight million eggs at a single spawning, although from four to six million is a more normal figure. As large numbers of mature cod assemble at the same period on the spawning grounds, astronomical numbers of eggs and vastly larger numbers of spermatozoa are shed more or less simultaneously. In this way the fertilisation of virtually all the eggs is assured. Despite the fact that the breeding fish are very particular in their choice of spawning grounds, this has no effect so far as the eggs and the fry that hatch from them are concerned. Like the eggs of nearly all our food fishes, the cod's egg is lighter than sea water and consequently, as soon as it is shed and fertilised, it rises into the surface layers, where it becomes a temporary member of the floating plankton, drifting at the mercy of the winds and currents. For a few days the young fish is developing within the egg, and is characterised at this stage by having a very slender body bearing a pair of relatively enormous eyes. It soon breaks out of the egg case and begins to feed on microscopic plankton organisms.

Floating free in the plankton the eggs and the young fry are of course very vulnerable, and enormous numbers of them are consumed by other plankton animals. When those that survive are about one inch long they seek shelter under the umbrellas of the giant jelly-fish *Cyanea arctica,* which in the early months of the year occur in enormous numbers throughout the sea areas in which the cod fry are found. They are known to the fishermen as sluthers, and are regarded with disfavour because fishing nets can easily become clogged with them. *Cyanea* is one of the largest species of jelly-fish, the umbrella often measuring a couple of yards across. They are carnivorous and catch quite large fish by means of long tentacles that hang down from beneath the umbrella. These tentacles are massed with stinging cells capable of paralysing instantly any fish which comes in contact with them. The astonishing thing is that the cod fry are apparently able to avoid the tentacles, and use the water between them as a haven with the certainty that no fish will venture between the tentacles to catch them. The fry of haddock and whiting are also found under the protective canopy of the sluthers.

After a week or two in this protective environment the cod fry are ready to sink to the bottom of the sea. By this time, through the movements of tides and currents, they have drifted a long way from the spawning grounds. Consequently, when they sink the majority of them are in the comparatively shallow water of creeks and estuaries, and for the first few years of their lives young cod occur mainly in these shallow inshore waters, only moving out into the open sea as they reach maturity.

The haddock, *Gadus aeglefinus,* is a much smaller member of the cod family, but its breeding habits are in the main similar. It becomes mature in its third or fourth year, and from then on undertakes annual spawning migrations to its chosen breeding grounds, in the same way as the cod does. The spawning season, during which each female lays about four hundred and fifty thousand eggs, lasts from mid-January until mid-June, reaching a peak in February and March. After hatching the young haddock, unlike young cod, remain in the surface plankton for about the first year of their lives, and have therefore grown to quite large fish when they finally settle on the sea bed.

As they approach maturity most fish undertake some sort of migration to particular areas to spawn. In some species these migrations are more spectacular than in others. By far the most astonishing of these breeding migrations is that undertaken by the common or European eel, *Anguilla vulgaris*, which is also known as the freshwater eel, although only part of its life is spent in fresh water.

For centuries man had been puzzled as to how the eels in our ponds and rivers reproduced themselves, as no breeding activity had ever been observed. In consequence all kinds of unlikely theories were advanced. 'Some say they breed, as worms do of mud,' says Isaac Walton in the *Compleat Angler*, 'and others say, that as pearls are made of glutinous dewdrops, which are condensed by the sun's heat in those countries, so eels are bred of a particular dew, falling in the months of May or June on the banks of some particular ponds or rivers.' According to other theories they grew from hairs, or were generated from turf; they were produced by black beetles, or even formed from particles scraped off the adults of their own kind.

The mystery of the eels' breeding activities was finally solved in the early 1920s by a distinguished Danish oceanographer, Dr Schmidt. Adult eels, which are yellowish in colour and sometimes known as yellow eels, are found not only in rivers and lakes but also in land-locked ponds all over Europe. They feed mostly at night on a diet of worms, crayfish and other small invertebrate life, lying up during the daytime under stones, in holes or under any cover they can find. Sometimes, when rain has been falling and the ground is wet, an eel may leave the water and travel overland to some other stretch of water. It is in this way that eels may appear in land-locked ponds where none have been known before.

For some years the eels continue to live in fresh water without changing, except in size, until they begin to approach maturity. Now their colour changes to silver during the summer, and with the approach of autumn these silver eels make for the sea, to begin the longest regular journey known to be undertaken by any fish. Travelling at a rate of some ten miles a day, they swim westward for week after week and month after month, until they arrive at last in the Sargasso Sea, having travelled well over two thousand miles across the Atlantic.

At last they have arrived at their breeding grounds, and at a depth of at least five hundred fathoms they spawn. No one has ever caught a spent adult eel returning from the Sargasso Sea, and it is believed that after this one incredible journey the adult eels die as soon as they have spawned. The accuracy with which these eels are able to cover such vast distances from their European rivers and arrive at their correct destination is most uncanny, and how they do it we cannot even guess. Even more astonishing, however, is how the tiny larvae that eventually hatch from the eggs are able without any guidance to undertake the return journey and arrive safely at the mouth of some European river.

So far as is known the minute eel larvae, not more than one quarter of an inch in length, begin to retrace the path by which their parents arrived as soon as they hatch in early spring. They are quite unlike eels in appearance, being flat and leaf-shaped. These curious little fishes, grown to about three inches in length, had long been known in European waters, but were not associated with eels. They were in fact regarded as a distinct species of fish to which the name *Leptocephalus* was given. Immense numbers of these *Leptocephalus* larvae set out on the eastward journey, but the majority fall prey to various predators long before they can reach their

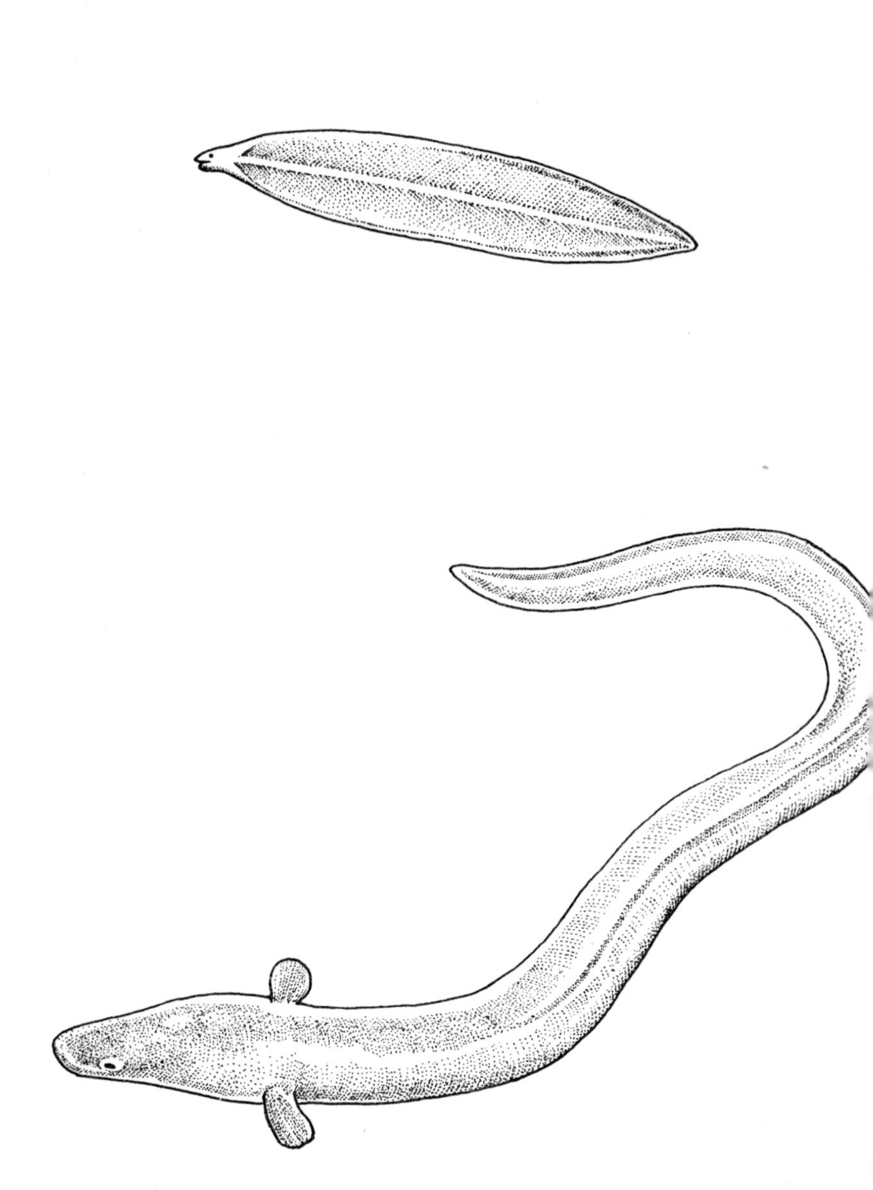

Leptocephalus larva and adult common eel (not to scale to each other)

ultimate destination. Those that survive to reach the comparative safety of European waters do so only after a continuous journey lasting for three years. In the autumn following their arrival they undergo a metamorphosis during which they acquire an eel-like shape. These are the young elvers, which now approach the coasts and swim up the rivers. Here, and in lakes and ponds, they will grow into adult eels. Eventually it will be their turn to undertake the great return journey to the place where they were hatched.

In the eastern parts of North America there is a very similar species, *Anguilla restrata,* which also spawns in the Atlantic in an area a little to the west of the spawning grounds of the common eel. Since the distance from the American rivers is not so great the little *Leptocephalus* larvae take only one year instead of three to cover the distance, and in adaptation to this shorter time they are ready to undergo metamorphosis and change to elvers at the end of this year.

Plaice and other flat fish adopt reproductive methods similar to those of the cod and haddock, shedding enormous numbers of reproductive products into the sea over selected spawning grounds where the eggs are fertilised and soon hatch to tiny larvae which drift about in the surface waters among the other animals of the plankton.

The chief interest in their reproduction is the amazing changes that take place during the first two months or so after the eggs have been fertilised. Each female plaice lays about a quarter of a million eggs, and the tiny larvae that hatch from them are so similar to the newly hatched larvae of typical round fish like cod and haddock that only an expert could tell the difference.

At four weeks old they have grown considerably and are about half an inch long, but they still look like any other fish larvae, with the pair of prominent eyes typical of all fish at this age. So far there is no hint of the dramatic transformation that will take place over the next four weeks.

But from now on the young plaice begins to develop abnormal proportions. Its body fills out but little, and it increases in length only slowly. Its depth, however, increases so rapidly that within another week it is already a noticeably flattened fish, though it is still swimming upright in the water. Further increase in depth over the next few weeks converts it to a really flat fish flattened from side to side.

At the same time an almost incredible change takes place. The left eye, until now situated on the left side of the head where you would expect to find any animal's left eye, begins to change its position. Slowly it travels upwards, turns across the top of the head and drops down over the right side. It comes to rest just above the right eye, the whole journey occupying about a fortnight.

The young fish, its body now fully flattened, no longer maintains its former upright position in the water, but swims with a definite list towards the now blind left side. Each succeeding day the list becomes more pronounced, until finally the fish is swimming right over on its side, with its two eyes facing upwards.

The side-to-side body and tail movement by which a round fish propels itself through the water becomes, in the flat fish, an up-and-down flapping of the body. The young plaice is now ready to abandon the surface waters and sinks down to the sea bed, where it will remain for the rest of its life.

Lying on its left side, the plaice uses mainly the left side of its mouth to obtain its food, and if you look at the head of a plaice you will see that the teeth in the left jaws are much better developed than those in the right.

The plaice is of course not the only flat fish. Dabs, flounders, witches, lemon soles, true soles and the halibut all develop in exactly the same way as the plaice, and finish up with both eyes on the right side. By way of variety, however, in the turbot and the brill it is the right eye that moves over to the left side, so that these fish lie on their right sides. And it is the teeth in the right side of the mouth that become more developed than those in the left side.

Although their larvae drift among the plankton in the surface layers of the sea, the adults of most of the important food fishes live and feed near the sea bed. They are described as demersal. By contrast the herring, *Clupea harengus*, is a pelagic fish, which swims in the surface waters in gigantic shoals containing millions of fish. For most of the year these shoals live in deeper waters well away from coasts, some probably at greater distances than others. The main reason for the annual migration which each shoal makes into shallow coastal waters is to visit its chosen breeding grounds to spawn. It is during these journeys to and from the breeding grounds that heavy catches are made.

It is not known what determines the beginning of the breeding migration of each shoal, but changes in water temperature, to which marine animals generally are very sensitive, may be an important controlling factor. In its spawning habits the herring is unique among our food fishes. All other species spawn into the sea, and the fertilised eggs, being slightly lighter than sea water, gradually rise to the surface layers. Here they drift until they hatch, as we have already seen, thus making a considerable contribution to the temporary plankton. The herring, however, produces eggs that are slightly heavier than sea water, depositing them in shallow water on the sea bed where it is stony and free from mud and weeds. The eggs remain in the crevices between the stones until they hatch.

The eggs of other fish drifting among the plankton receive no protection, and a large proportion are eaten by fish and other marine animals before they have had time to hatch. But among the stones the herring eggs do get a certain amount of protection, though haddock are particularly fond of them. In fact many of the herring breeding grounds were first located through catches of 'spawny' haddock—that is haddock whose stomachs were full of herring eggs.

An interesting theory has been put forward to account for the unusual breeding habits of the herring and for its breeding migrations. It is suggested that the herring originally lived for most of the time in the sea, but travelled up rivers to spawn, depositing its eggs on stony river beds, just as the salmon does today. Eggs laid in fast-moving rivers would have to be fixed to the river bed to prevent their being swept out to sea. At this time Britain was part of the continent of Europe, and the present North Sea bed was above sea level, forming a plain across which the Rhine flowed northwards to enter the sea somewhere to the north of Scotland. It would be up this and other smaller rivers flowing in a similar direction that the herring would swim on their spawning journeys. When the vast plain became a shallow sea these spawning journeys persisted as migrations from deeper to shallower water, and the habit of fixing the eggs to stony ground was retained.

One of the most unusual examples of fish breeding behaviour is that shown by the grunion, *Leuresthes tenuis*, a small inshore fish which lives in Californian coastal waters. In its breeding activity it is dependent upon the moon. The breeding season extends from March to June, but actual breeding occurs only at precise times over this four-month period. On the second, third and fourth nights following each full moon, and just as the tide is on the turn, having reached its highest point up the beach, the grunion come in in pairs. Once they have reached the very edge of the water the females wriggle into the sand tail first until at least half of their bodies are buried. Then they lay their eggs. Meanwhile the males have curled around the females, and deposit their sperm over the eggs to fertilise them.

After the fourth night the tides no longer reach the top of the beach, so the eggs can develop undisturbed in the sand. By the time the next spring tides are due a fortnight later the eggs are ready to hatch. This they do at high tide, and as the young fish emerge on to the surface they find water in which they can swim down the beach to reach the offshore water where they can continue their development. If they emerged earlier they would find themselves stranded high and dry, and would rapidly perish.

There is some evidence to show that the eggs are in fact ready to hatch about a week after being laid, but will only do so under the influence of the shaking produced by the high tide waves. At least, fertilised eggs kept in a laboratory aquarium will not hatch until they have been shaken, even though pre-hatching development has been completed long before.

In contrast to the majority of marine teleosts, which show no concern for their offspring, those that live on the shore in the zone between the high and low tide marks show a considerable amount of parental care, and in consequence produce vastly fewer eggs. Many of them guard their eggs until they hatch, which not only protects the eggs from potential enemies, but also ensures that they hatch in a habitat suitable for their survival. If they were allowed to be swept out to sea, the young fish hatching out might never be able to regain the shore.

Not only do the shore fish practise parental care, which is an unusual feature in the bony fish, but most of them exhibit it in an unusual form. It is usually the male who looks after the developing eggs, the female who swims away and takes no further interest in them once they have been laid. The female common blenny, for example, lays her eggs during the summer in rock crevices or under rock ledges and then swims away, leaving the male to stand guard over them, prepared to attack any other fish that comes near.

Another blenny, the gunnel or butterfish, is one of the few shore fish in which the female does share in the guarding of the eggs, taking it in turn with the male to lie with her long body coiled around them. A third member of the blenny tribe that deserves a mention is the viviparous blenny, one of the few marine teleosts to retain the eggs within the female body until they hatch, which they do in about three weeks, but the young fish are then retained within the female oviducts for a further two to three months until they are well developed. When they finally emerge they are up to an inch and a half in length.

Many of the shore fish have developed special means for attaching themselves to the rocks so that they may not be swept away in heavy seas. In these the pelvic fins unite beneath the fish to form a kind of ventral sucker. The little gobies, among the most common of the shore fishes, show this modification. They are small fish and rather inconspicuous, their spotted and banded bodies often blending well with their background. Their eyes are relatively large, and have shifted almost to the top of the short head, so that they look upward and outward as the little fish adhere to the sides and bottoms of the rock pools where they are usually found.

At breeding time it is the male that takes the initiative, first searching for a suitable sheltered stone or rock surface where the eggs can be laid. He then goes off to search for, and often to fight for, a mate, who lays her eggs on the chosen stone or rock, the male fertilising them as they are laid. This done, she goes off again leaving the male to guard them until they hatch. During their development he sends a steady current of water over them by the movements of his fins. As soon as they hatch the male abandons the brood and goes off in search of another partner. Before the summer comes to an end each of the original pair will probably have been responsible for several broods with a variety of different partners.

The idea of fish constructing a nest may seem rather peculiar, but there are two kinds of seashore fishes that do just this. The better known of these nest-builders are the sticklebacks. At the breeding season the males construct nests by binding together living seaweed fronds with the aid of thread-like secretions produced from their kidneys. These nests are pear-shaped and about four inches in length, with a cavity in the centre produced by a boring action of the head of the fish. When he has made his nest ready the male goes out to search for a mate. After a quite elaborate courtship dance in the water she is finally persuaded to enter the nest to lay her eggs, after which she departs and the male enters the nest to fertilise them. He then mounts guard over them, forcing a continuous current of water through the nest to aerate the eggs by vigorous movements of his pectoral fins. Any other fish that approaches is attacked and driven off.

The various species of seashore wrasses also make nests for the reception of their eggs, but these are less elaborate than those constructed by the sticklebacks. They are in fact merely masses of severed seaweed fronds wedged into rock crevices, with fertilised eggs deposited among them. There is no actual parental care.

Although not common, hermaphrodites do occur among several groups of marine teleosts, notably the sea perches and the sea breams. In some of the species one part of the gonad produces male reproductive products while the other part produces eggs. Self fertilisation may be avoided by one part of the gonad ripening before the other. In one family of sea perches the gonad is not divided into two functional parts. Instead the younger mature fish are males, and as they get older they turn into females by a reversal in the structure and function of the gonads. In yet other species ripe spermatozoa and eggs are produced at the same time, so that self fertilisation can occur.

Viviparity is another phenomenon that is not very common among marine teleosts. Perhaps the best known viviparous species is the little guppy, *Lebistes reticulatus,* in which after internal mating the eggs are fertilised while still enclosed within their follicles in the ovary. After an initial period of development the follicles become ruptured and the eggs are released into the cavity of the ovary. They are then nourished by secretions produced by special follicular cells until they are sufficiently well developed to be born.

Lebistes is really ovoviviparous, but there are other teleosts in which there is true viviparity, involving some kind of placenta. In the dwarf top-minnow, *Heterandria formosa,* for example, the walls of each ovarian follicle develop a surface of blood capillaries extending into villi which make intimate contact with the surface of the developing embryo, and thus supply it with all its needs as well as carrying away all its waste products. After mating *Heterandria* is able to retain any spermatozoa unused in its ovary for upward of a year, and is thus able to fertilise successive batches of eggs with only one mating. Retention of living spermatozoa also makes it possible for two or more broods of different ages to be developing at the same time in the ovary.

'Mouth breeding' is another device used by certain teleosts to protect their developing embryos. Among the marine teleosts exhibiting this phenomenon are some of the cat-fishes and the cardinal fishes, but it is more common among freshwater species. *Galeichthys felis* is a cat-fish that lives in American coastal waters. It lays exceptionally large eggs, up to twenty millimetres in diameter. After they have been laid and fertilised the male swims around collecting them up in his mouth, where he is capable of storing as many as forty-five. Here they remain for about four weeks before finally hatching. Even then the male's task is not completed, for the young fish remain in his mouth for further protection for another couple of weeks or so. Many of the cardinal fish produce very small eggs, enabling the males of certain species to house as many as several hundred eggs in their mouths. The record, though, seems to be held by a Mediterranean species, *Apogon imberbis*, in which the male is credited with being able to find room for an astonishing twenty thousand minute eggs.

Mention of eggs developing in a special abdominal pouch immediately calls to mind the kangaroos and wallabies. But one group of fishes, comprising the sea-horses and the pipefishes, also incubates its eggs in pouches, though it is the male and not the female who possesses the pouch and supervises the incubation.

The sea-horse is one of the most unusual of all fishes. The head really does resemble that of a horse, or the knight in a chess set. Unlike almost all other fishes it swims in an upright position, and the tail, which is used normally to propel a fish through the water, in the sea-horse has lost this function and become prehensile, the little fish using it to attach itself to weeds. Of the paired fins only the front ones, the pectorals, are represented, and these serve to maintain the vertical position of the fish. The only fin capable of moving the fish is the dorsal fin, which when in action oscillates rapidly and looks rather like a propeller. But the sea-horse does not swim much, remaining motionless awaiting tiny animals that are its prey to swim in the vicinity of its long suctorial mouth.

The brood pouch of the male is located at the base of the tail and envelops the opening of his sperm duct. After a preliminary courtship the female lays her two hundred or so eggs in the pouch, and as she does so they are fertilised by sperm shed over them from the sperm ducts. Once the eggs have entered the pouch the wall of the pouch develops a spongy consistency, with a vast increase in blood capillaries, and into this spongy tissue the eggs become engulfed. Soon an intimate contact is established between the developing embryos and the spongy wall of the pouch, constituting a placenta through which they are nourished.

The young sea-horses are sufficiently developed to leave the paternal pouch in about three months. In warmer seas the male may acquire a second brood within a day or two of discharging his first, and in exceptional cases he may rear three broods in a single season.

Some pipe-fishes have brood pouches similar to those of the sea-horses, but in other species there is no brood pouch, the male merely attaching the fertilised eggs to the underside of the tail. There is of course no kind of placenta, the eggs having to rely upon their stores of yolk to provide them with the food materials needed for their development.

The majority of freshwater fishes are not particularly distinguished in their breeding habits, merely laying their eggs on weeds or on the bed of the river, lake or pond in which they live, and leaving them to take their chance with whatever predators may be around.

Salmon and trout, however, are an interesting group, quite apart from their value as luxury food. The common trout, *Salmo trutta*, exhibits remarkable variability both of form and habit. The typical form is known as the brown trout, and is widely distributed in lakes and rivers. In these the conditions vary considerably. In some there may be an abundance of food, whereas in others the fish are hard put to it to make a living. Conditions of acidity also vary, and between them these two factors exercise a considerable influence on the growth rate and the ultimate size of the fish. In some rivers and lakes brown trout up to two feet in length are not uncommon, whereas in others they seldom exceed a few inches in length.

The brown trout is widely distributed throughout Europe and is also found in Iceland. In Britain and other temperate places breeding occurs during the winter. Spawning takes place in fast moving streams where the bed consists of clean gravel. The females select the breeding sites, and excavate hollows known as redds by sweeping stones away with their tails. While they are doing this the males appear, selecting their mates and driving potential rivals away. As soon as the females lay their eggs in these redds and they are fertilised by the males, the females cover them by sweeping gravel over them.

The female trout lays an average of about one thousand eggs for each pound of her body weight. As they develop they are oxygenated by the clear water flowing through the stones under which they have been buried. Incubation takes between two and three months, but even after they have hatched the tiny larvae remain among the stones for a further period of two or three months. When they do finally emerge from their shelter their yolk sacs have been completely absorbed and they have acquired the typical streamlined form of the adult trout.

Quite different in colour from the typical brown of the river and lake forms of *Salmo trutta* is the silver of the migratory form, which is known as the sea trout or salmon trout. For a long time it was classified as a separate species. Like the brown trout, salmon trout breed in fast-moving streams, burying their eggs in the gravel of the stream bed. The young fish remain in the rivers until they are about two years old, when they swim downstream and out to sea. It is then that they develop the silver colour that distinguishes them from the non-migratory brown trout. After another year or more, during which they have become mature, the salmon trout return to the rivers to spawn.

Despite these excursions of some members of the species to

the sea, *Salmo trutta* must be regarded essentially as a fresh-water species, whereas the related salmon, *Salmo salar*, is much more at home at sea. Not only does it spend most of its adult life at sea, but most of its feeding is done there also. Only the young fish, before they leave the rivers for the first time, feed in fresh water. When they return as adults to spawn they cease feeding as they enter the rivers and will not feed again until they return to sea after spawning, which may be many months later. By this time they are usually in very poor condition.

Their phenomenal ability to overcome obstacles such as waterfalls and weirs as they ascend a river on their spawning runs is well known. They make for the upper reaches where the water is shallow, clean and swift-moving, and the bed is made of clean gravel. Here the mature fish excavate shallow depressions by lying on their sides and lashing with their tails. Into these redds the females lay their eggs and the males cover them with sperm to fertilise them before the parents sweep the gravel back to cover them.

The tiny fry that hatch from these eggs are known as alevins, and they remain lurking among the stones of the redds for protection for some weeks. When they are about an inch long they leave the redds as parr, which are distinguished by about ten dark oval marks arranged in a row along each side of the body. When they are about two years old the parr, now some five to six inches long, change to smolts, losing their parr markings and becoming silver in colour. These smolts represent the first migratory stage, and they move down the rivers and out to sea, where they will remain for the next few years until they have become sexually mature. A few do in fact become mature after one year, returning as grilse to the rivers to spawn, and back again to the sea as soon as they have done so.

The principal food of the salmon at sea is fish, especially sand eels, herring and mackerel. Spawning is a very exhausting business for the salmon, production of the great quantities of spawn being a severe drain on its resources. Many fish are believed to manage only one spawning in their lives, and few probably manage more than two or three.

Salmo salar is a widely distributed fish on both sides of the North Atlantic, ascending almost every unpolluted river of north-western Europe and eastern North America. The salmon of the Pacific, which ascend the rivers of western North America, and which form the basis of the huge canning industry, belong to a different genus. The best known species are the humpback or pink salmon, *Oncorhynchus gorbuscha,* the chinook or king salmon, *Oncorhynchus tschawytscha,* and the sockeye or red salmon, *Oncorhynchus nerka.* These species are more completely marine than any species of *Salmo,* leaving the rivers when they are quite young and returning only briefly at the breeding season to spawn.

Somewhat similar to the brown trout in their spawning habits are the tiny bullheads, which are found in cool fast-flowing streams in most parts of Europe. They average between three and four inches in length, and spawning time is in the spring. At this time the male excavates a nest in the gravel on the river bed, and into this the female deposits her few hundred eggs, to be fertilised by the male. He now mounts guard over them until they hatch in three to four weeks, fanning them with his large pectoral fins to keep them well aerated. Once they have hatched he takes no further interest in them, but until they have absorbed their rather large yolk sacs they remain among the stones for protection. Even the adults remain on or near the river bed, feeding upon such organisms as can be found there.

In contrast to the trout and the bullheads, in which parental care drastically reduces the number of eggs each female must produce to keep the species going, the perch makes no provision for the protection of its eggs. In consequence each female lays something like one hundred thousand eggs per pound of her body weight. These are laid in strings attached to water plants or any inanimate objects projecting from the bed of the river, and here of course they receive no protection from potential predators.

The greatest single difference between marine and freshwater teleosts so far as their reproduction is concerned is that, whereas the vast majority of marine species lay enormous numbers of buoyant eggs that float near the surface of the sea in the plankton, the majority of freshwater species lay eggs that are heavier than the water, so that even if they are not adhesive, or buried in the river bed, or attached to water plants they still sink to the bottom where they cannot be swept away by the fast-moving currents above them.

Besides salmon and trout a number of other teleosts bury their eggs in gravel. Perhaps the most interesting of these is a group of annual fish known as top-minnows, which occur in rivers in certain parts of both Africa and South America, where a rainy season alternates with an excessively dry season when smaller streams become reduced to caked mud. During the rainy period the top-minnows mate and bury their fertilised eggs in the muddy beds of these streams. When the streams dry up during the dry season all the adult fish of course die, but the eggs, which have been provided with a tough resistant coating, survive in a dormant state until the next rainy season, when they hatch out to produce the top-minnow population for that year.

As with marine teleosts, there are a number of specialised adaptations among the freshwater forms, most of which parallel those found among the marine species. Nest-building is indulged in by quite a number of freshwater teleosts, some of them, including the freshwater sticklebacks, constructing a nest of weed similar to that of the marine stickleback. But the most unusual nests are the bubble nests constructed by quite a number of species that live in stagnant tropical waters. The bubbles are produced by a pair of special organs situated in the gill chambers. Air passing through these labryinthine organs, as they are called, form bubbles with which the nest is carefully constructed. The eggs are laid and fertilised in these bubble rafts, where, developing close to the surface, they are sure of an adequate supply of oxygen. Some of the better known of these bubble nesters are the Siamese fighting fish,

Betta splendens, some of the gouramis, and the paradise fish, *Macropodus opercularis*.

Mouth breeding is also practised by some freshwater species. *Tilapia mossambica*, an African cichlid, produces between twenty and fifty eggs. As soon as these have been fertilised the female gathers them up in her mouth. While they develop she retires to some kind of shelter to rest. When the eggs hatch in about twelve days the larvae still remain around their mother, following her in a shoal as she moves about. If danger threatens she is apparently able to call them to her mouth, where she can give them maximum protection. For some reason as yet unknown the eggs will not develop if they are removed from their mother's mouth: they soon become mouldy and die.

The male and female of *Betta brederi*, an Indian relative of the Siamese fighting fish, play quite a game with their eggs. As they are shed by the female the male catches them in his anal fin, and while they are cradled there he fertilises them. He now releases the eggs, and as they float gently down through the water the female hurriedly swims around gathering them up in her mouth. But she has no intention of incubating them. This is the male's job, as it is in most mouth breeders. Once she has gathered up the whole clutch she swims into a position facing the male, when she spits the eggs at him one by one, and he catches them and stores them away in his mouth, where they will remain until they hatch.

Although the lung fishes today comprise a group that has become almost completely extinct, they are extremely important because of the position they occupy in the evolution of the land vertebrates. At the same time as the first actinopterygians and elasmobranchs were being differentiated from their placoderm ancestors, another group of placoderms was giving rise to the first of the crossopterygians or lung fishes. During the Devonian period three distinct orders were developed.

The earliest of these were the osteolepids, long, heavily built fishes with a body covered with thick scales like those of the paleoniscids. The most important external feature, however, was that the paired fins, instead of being wide and winglike, were modified to form long thin appendages rather like legs without feet. By the Carboniferous period these osteolepids were becoming rare, and by the early part of the Permian they had become completely extinct. They left behind them, however, a group of descendants that formed the second order, the coelacanths. Until 1938, when some fishermen landed a specimen of an undoubted coelacanth caught off the coast of East Africa, it had been believed that this group, too, had become extinct by the end of the Cretaceous period, that is some sixty million years ago.

Interesting though this discovery of *Latimeria* was, from the point of view of the evolution of the tetrapods or land vertebrates the survivors of the third order, the Dipnoi, are of more importance. Like the other two orders, the first members appeared in the Devonian, some three hundred million years ago. In general appearance they were similar to the members of the other two orders, having their deep bodies heavily armoured with large thick scales, and lobed, limb-like paired fins. Three kinds have survived to the present time, but they show considerable modifications to the original plan. The scales have become thin, though still large and prominent, and there has been a considerable reduction of bone in the skeleton.

Of these three surviving species the most primitive and probably the oldest is the Australian lung fish, *Neoceratodus forsteri*. Its fins are still leaf-shaped like those of the early Dipnoi, and it has a deep body covered with large scales. It is also the largest of the three species, growing to a maximum length of six feet and a weight approaching one hundredweight. It is neither common nor widespread, occurring in fact only in two Queensland rivers, the Burnett river and the Mary river. It is probably nearer in general structure to the ancestral lung fishes from which the amphibia were originally evolved than are the other two surviving species.

The African lung fish, *Protopterus annectans*, is widely

distributed in the rivers of tropical Africa, while the South American lung fish, *Lepidosiren paradoxa,* is found in many of the rivers in the Amazon region. In both the body is less heavily built than in *Neoceratodus,* and the scales are thinner. The fins, too, are no longer leaf-shaped, but have become reduced to thin filaments which are very long in *Protopterus.*

In view of the evolution of the amphibia from some ancestral crossopterygian, the breeding habits of the three living lung fishes are particularly interesting. *Neoceratodus* lays large eggs surrounded with a considerable layer of jelly, thus making them similar in appearance to typical amphibian eggs. The female deposits them either on the bed of the river or among aquatic plants.

Protopterus and *Lepidosiren* have a more elaborate method of breeding. As the breeding season approaches the males excavate nests in the mud on the river beds. *Protopterus* nests are large round holes, over which the male mounts guard after the eggs have been laid in it and fertilised. By continuous movements of his tail he keeps the water circulating over the eggs and thus helps to aerate them. The *Lepidosiren* nest is a burrow which may be as much as four feet in length. In both species the larvae which hatch from the eggs are quite unlike the adult fish, but bear a striking resemblance to amphibian tadpoles, both in general shape and in the possession of external gills and suckers beneath their heads.

CHAPTER THREE

AMPHIBIANS

The amphibians—the frogs and toads, the newts and salamanders—may be today a relatively insignificant group of animals. Their importance is due to their great evolutionary significance rather than to their present or past dominance or abundance—they have in fact never been either dominant or particularly abundant. But from the evolutionary point of view they hold an extremely important place in vertebrate development. They provided the bridge across which the aquatic fishes were able to give rise to the completely terrestrial reptiles, which in their turn were the ancestors of the birds and mammals and therefore of man himself. Their methods of reproduction particularly reflect their intermediate position.

To produce a completely terrestrial animal from a completely aquatic ancestor four major changes were necessary. The fins of a fish would be of little use for moving about on land, so they had to be converted into pairs of limbs, the pentadactyl limbs which are typical of all terrestrial vertebrates.

Even if a fish could move about on land it would not be able to survive for long because it would be unable to breathe. Its gills are designed to extract oxygen from the water in which it is dissolved, and are incapable of extracting it from the air. For this purpose a lung is necessary. The amphibian lung, however, is quite a primitive one, in most species no more advanced than the simple lung of the lung fishes, and is thus not very efficient. The important thing from the evolutionary point of view, however, is that the amphibia did develop a pair of lungs whose efficiency could be increased in the higher vertebrates.

Another reason why a fish would be unable to live out of water indefinitely is that water would be lost through its skin, so that death would occur as a result of desiccation. The reptiles, birds and mammals have developed a skin that prevents uncontrolled loss of water, but the amphibian skin has not acquired this power. As a result amphibians cannot survive under drying conditions, and are therefore mainly confined to damp and shady places. Their moist skin, however, serves to supplement their very inefficient lungs. It is extremely well supplied with blood vessels and these, together with its moist surface, enable it to extract oxygen from the air. In the majority of amphibians in fact the skin is a good deal more efficient than the lungs, and in some species the lungs have degenerated completely.

It is in their reproduction, however, that the amphibians are least emancipated from water. The majority of fish shed their eggs into the water, so that, whatever requirements they may have, prevention of desiccation is not one of them. Reptile and bird eggs do not dry up because before they are laid they are covered with a protective shell. Amphibian eggs, however, have no such protective covering, and have to be laid in water. At the breeding season, therefore, however well adapted amphibians may be to living on land, the vast majority of them must make their way back to the water to breed. From their eggs hatch the familiar amphibian tadpoles which are completely adapted to an aquatic existence, breathing by means of gills similar to those of a fish, and swimming by means of a muscular, fish-like tail. Only after a prolonged period of development lasting for several months do these aquatic tadpoles undergo an extensive metamorphosis which converts them to tiny land animals. At this time lungs replace their gills, two pairs of limbs develop, the skin becomes converted to a highly vascularised respiratory surface, and in the case of the frogs and toads the tadpole tail is completely absorbed.

The Amphibia form one of the eight classes comprising the subphylum Vertebrata. Three subclasses are recognised, the subclass Anura or Salientia, consisting of the frogs and toads, the subclass Urodela or Caudata, consisting of the newts and salamanders and their allies, and the subclass Apoda, a group of relatively little known worm-like amphibians without limbs.

Frogs and toads are less dependent upon water than newts and salamanders, except of course at the breeding season, and are much more specialised in their adult structure. After mating and egg-laying in the spring or early summer the adult frogs and toads are free to leave the water and go where they will. During the remainder of the summer they may roam some distance from the breeding pond. As winter approaches they will turn their attention to hibernation. Frogs usually choose to hibernate in the mud at the bottom of a pond, but toads will hibernate almost anywhere where they can be sure of cover through the winter months.

When they finally wake up in the early part of the year they may thus be a long way from the pond in which they bred the previous year, or in the case of young frogs and toads the pond in which they were hatched and developed. Now, however, they will set off in search of their home pond, even passing by and completely ignoring other apparently equally suitable stretches of water. How this return to the home pond is achieved is not known, but R. M. Savage, who devoted a great deal of time to investigating the phenomenon, put forward an interesting theory. In a particular area which he studied intensively there were fifteen apparently similar ponds, but only certain of these were ever used by breeding frogs and toads. He believed that it was the particular 'smell' of the pond which attracted the frog or toad to it, and that this smell was caused by the essential oils produced by the plants living in the pond. Whatever the cause, the reality of homing is beyond dispute, and in general toads are more successful than frogs in returning to their home ponds to breed.

The first arrivals are usually males. When the females start to arrive they are picked out by the males, each climbing on to the black of a female and gripping her so firmly with his front legs that the two cannot be physically separated without serious risk of damage to the male. His arms pass just behind her armpits and meet or nearly meet across her chest. The pairs remain in amplexus, as this mating embrace is called, until all the eggs have been laid and fertilised. As the breeding season gets under way and members of both sexes begin to converge on the breeding pond any male who meets up with a female before reaching the pond will usually jump on to her back and ride the rest of the way, so that many of the later arrivals are already in amplexus as they dive into the water.

Quite fierce fighting can take place between male frogs and toads for possession of the females. The male already in possession uses his very powerful hind legs in defence, while the attacker tries to prise open his arm grip upon the female, but however frenzied the attack it is seldom successful. The urge for unmated males to clasp may lead them to grasp at any moving object. Occasionally fish have been seized just behind the head. This closes their gills so that they cannot breathe, and so they drown.

Whatever else may influence their choice of a pond, frogs do certainly prefer ponds that have water flowing through them. The actual spawning site is usually in shallow water and exposed to the sun. In most ponds there is only one spawning site used by all the pairs coming to the pond, although a large pond may have two, and very occasionally three. When a large number of pairs is involved the resulting spawn may cover between twenty and thirty square yards.

Females of medium-sized frogs such as the common European frog, *Rana temporaria*, or the edible frog, *Rana esculenta*, lay up to three or four thousand eggs at an average spawning. These are all discharged at once over a period of a few seconds, and as they emerge the male sheds his spermatozoa over them to fertilise them. Immediately afterwards he relinquishes his hold on his mate and swims off.

As they pass down the oviduct of the female before they are laid the eggs of the frog are coated with a thin layer of very concentrated jelly, and this absorbs water rapidly as soon as they are discharged. As a result each egg, measuring between two and three millimetres in diameter, comes to lie in the centre of a sphere of jelly up to ten millimetres in diameter. When first discharged and fertilised the eggs sink to the bottom of the pond, but they rise to near the top as soon as the water has been absorbed. This diluted jelly is a remarkable substance, since it retains the properties of a solid yet consists of 99.7 per cent of water. Not only does it give physical protection to the egg during its initial development, but it also enables it to develop at a temperature significantly higher than the cold water of the pond. In temperate climates pond water is still very cold in the early months of the year when frogs spawn. But the clear jelly allows the sun's rays to reach the eggs, which being dark absorb heat from the rays, and this heat is retained by them because the jelly acts as an insulator. The difference in temperature between the developing eggs and the surrounding water may be as much as two degrees centigrade, and this difference can have a very noticeable effect on their rate of development.

Toad spawn can easily be distinguished from frog spawn because the eggs are laid in a continuous string contained in a uniform cylinder of jelly. Egg-laying is not an instantaneous process as it is in frogs, but takes place intermittently over a period of several hours, each batch of eggs being covered with an emission of spermatozoa as they emerge. Localised movements of the mating pair during this time serve to twist the egg string around aquatic plants, where they will remain anchored until the eggs hatch. The egg strings usually measure between seven and ten feet in length, and contain between three and five thousand eggs.

Members of the family Bufonidae possess a peculiar additional reproductive organ known as Bidder's organ. This family contains many of the best known of the world's toads, and one or more of its hundred or so species can be found in almost every part of the world. Bidder's organ is present in both sexes, but is better developed in the male than in the female. It is attached to the front of the ovaries or testes, and in both sexes consists of immature egg cells. If the true reproductive organs are removed Bidder's organ becomes functional, producing mature eggs which, suitably fertilised, can produce tadpoles.

Frogs and toads may not have many obvious uses, yet in recent times they have invaded the field of medicine. In 1934 the African clawed toad, *Xenopus laevis*, broke into the headlines when it was discovered that the female of the species could be used as a test for human pregnancy. This was an important discovery, because for centuries man had searched with little success for a reliable and quick pregnancy test. Unlike most female frogs and toads, the female clawed toad carries her eggs throughout the year, finally expelling them when she eventually mates. Now it was found that she would also do so if she was injected with one of the female hormones which is produced in large quantities in the early stages of human pregnancy, and which is present at this time in the urine. A minute amount of this urine injected into the female clawed toad stimulates it to begin expelling its eggs within a few hours.

More recently, in 1947, another equally reliable test was devised in which the males of any species of frog or toad can be used. In this case injection of human urine containing the female hormone causes the frog to extrude spermatozoa within about two hours. This method has the advantage that local species of frogs and toads can be used, thus making it unnecessary to import toads at considerable expense from Africa.

The balance of nature often becomes apparent only when it is disturbed, and unforeseen consequences result. After 1934 so many clawed toads were collected and exported from various parts of Africa that their numbers were decreased very considerably. As a result mosquitoes became much more plentiful, to the annoyance of the local residents, who not unreasonably complained. Clearly the clawed toad was a much more important factor in keeping the mosquito population within bounds than had hitherto been realised.

As we have already seen in fishes, any kind of protection, parental or otherwise, in the early stages of development reduces the total number of offspring it is necessary to produce, because it is in the early stages that any animal is most vulnerable to attack by its enemies. A number of quite unrelated species of frogs and toads have evolved methods of protecting their eggs, and in some cases the resulting tadpoles as well, from such attacks.

One of the best known of these is the European midwife toad, *Alytes obstetricans.* The female produces her eggs in a long string which may be as much as four feet in length and contain sixty to seventy eggs. As they emerge they are fertilised in the usual manner by her mate, who then proceeds to gather them up and twist them in a bunch around his hind legs. For three or four weeks they must be a great handicap to him as he carries them about, often having to go for a swim to keep them moist, because if he let them get too dry they would die. Then, when they are due to hatch, he returns to the pond and stays there until all the now well developed tadpoles have wriggled out of the eggs and swum away free into the water, leaving him to discard the string of empty egg cases.

In the pouched or marsupial frogs of Central America, belonging to the genus *Nototrema,* it is the female that looks after the fertilised eggs. Only a dozen or so are laid, but these are considerably larger than normal frog eggs because they contain a much greater quantity of yolk. The females possess a fold of skin in the form of a pouch across the hind end of the back, and as soon as they have been fertilised the eggs are pushed into these pouches by the males using their hind feet. Even when they have hatched the tadpoles do not leave the safety of the pouch, but remain there until they have achieved metamorphosis, only emerging to assume an independent existence when they have developed into young frogs. It is to provide the food needed for this prolonged development that the eggs are stocked with their unusually large amounts of yolk. When they are first hatched the tadpoles are provided with a peculiar bell-shaped structure which covers and protects their two pairs of external gills.

Another somewhat similar method of maternal care has been developed in the Surinam toad, *Pipa americana,* a common inhabitant of the damp South American forests from Guiana as far south as Brazil. At the onset of the breeding season the skin on the back of the female toad becomes soft, thickened and glandular. She then lays fifty or so eggs into the water, and after the male has fertilised them he proceeds to take them up one by one and press them into this soft skin, which soon closes over the depression. Thus each egg finds itself enclosed in a tiny pocket filled with fluid where it is completely protected from any potential enemies. Here the tadpole undergoes its full development, including its metamorphosis into a minute young toad. Only then does the little toad force its way out of its tiny nursery to begin its independent life. Because of the confined space in which it has had to develop it is much smaller than the newly-metamorphosed tadpoles of other frogs and toads of similar size. As soon as the tadpoles have emerged the mother rids herself of the additional superficial layer of skin by rubbing it off against stones or plants. At no time during its development does the Surinam toad tadpole develop gills; respiration is effected through its skin.

One of the most remarkable of these reproductive adaptations designed to provide parental care for the developing tadpoles is shown by Darwin's frog, *Rhinoderma darwinii,* a small Chilean species of tree frog first discovered by Charles Darwin. As the breeding season approaches the frogs leave their trees and migrate to the nearest stretch of water. Here the females lay a dozen or so eggs and the males fertilise them. As soon as this has been done the males then swim around gathering them up one by one and apparently swallowing them. They are not cannibals, however. It falls to their lot to look after the eggs until they hatch, and they pass, not into the stomach but into the vocal sacs at the back of the mouth, which are specially enlarged and modified as brood pouches to accommodate them. Here the eggs remain immersed in water, soon hatching to tiny tadpoles. They do not now emerge, but remain in their unusual nursery for another two or three months until they have completed their metamorphosis into tiny froglets. Only then do they emerge to face the hazards of the outside world.

True viviparity, in which the fertilised eggs are retained and undergo their development within the oviduct of the female, has been developed in only two rather rare East African species of frogs, *Nectophrynoides tornieri* and *N. vivipara*. The tadpoles develop an unusually long tail, the surface of which is well supplied with blood vessels. The blood in these vessels absorbs food and oxygen from equally abundant blood vessels in the wall of the oviduct, with which the tadpole's tail maintains close contact.

Guarding their eggs either in or on their own bodies is not the only method adopted by frogs and toads to protect their developing offspring. A number of different methods of protecting them without guarding them have also been evolved. One of the most elaborate of these is used by the blacksmith tree frog, *Hyla faber*, from Brazil. Before she is due to lay her eggs the female seeks out a pond, and at the very edge, where the water is only an inch or two deep, she scoops out a hollow basin. Around it she throws up a bank of mud separating this tiny artificial pond from the main pond, patting down the mud with her feet to make it firm. Into this pond she lays her eggs, which are fertilised by her attendant mate and then left to develop in comparative safety, the mud bank shielding them from predators living in the big pond.

Dr Goeldi first investigated this remarkable achievement, and in 1894 was able to witness the complete construction of a nesting pond.

> We saw a mass of mud rising to the surface [he wrote in his report], carried by a tree frog, of which no more than two hands emerged. Diving again, after a moment's time, the

frog brought up a second mass of mud, near the first. This was repeated many times, the result being the gradual erection of a circular wall. From time to time the head and front part of the body of the builder appeared suddenly with a load of mud at some point; but what astonished us in the highest degree was the manner in which the frog used its hand for smoothing the mud wall, as would a mason with his trowel. And by examining the hands of this tree frog it will readily be understood how they are most serviceable trowels, their terminal joints bearing large expansions. This careful process of smoothing could be better observed as the wall gradually heightened, until it reached about four inches, when the frog was compelled to come out of the water.

The parapet of the wall receives the most careful smoothening, the outside being neglected, and the levelling of the bottom attained by the action of the creature's body, aided by the hands. The aspect of the pool may be compared to the crater of a volcano, or a vessel of a foot in diameter filled with water. Although the female undertakes the entire task of building, she is incommoded the whole time by the male sitting on her back. Should he be frightened from his post, he will soon emerge from the water at a distance of a few feet, when, if signs of danger be wanting, he will climb the walls of the nest and regain his original seat.

In the Gran Chaco region of Northern and Central Argentina there are great areas of marshy swamp. In the summer the water becomes so warm that much of its oxygen is driven out, and there would be insufficient oxygen for the proper development of frog eggs. As a result the frogs that live in these regions, mainly *Physalaemus* and *Leptodactylus* species, have developed a method of overcoming this difficulty. As the female lays her eggs and the male mounted on her back covers them with seminal fluid to fertilise them he also beats his hind legs rapidly. This results in the mucus that covers the eggs being whipped up into a froth to form a foam nest which floats on the surface of the water. Within this nest the eggs are kept moist and at the same time are able to obtain sufficient oxygen. These floating nests, which look very much like snowballs, are a conspicuous feature of the Gran Chaco swamps.

A number of different tree frogs adopt a similar method, but their foam nests are fixed to the leaves of trees overhanging water. The foam from which these nests are made sets to a hard meringue-like consistency, and within it the developing eggs are kept cool and moist. It is not durable, however, and before long it begins to decay. By the time the tadpoles hatch it has become quite rotten, so that they are able to fall through it and drop into the water below, where their further development takes place. One Malayan species that uses this method is *Rhacophorus malabaricus,* one of a group known as flying frogs because the webs of all four feet are so well developed that they are said to be able to glide for short distances.

One American species of leaf frog, the Antillian frog, *Hylodes martinicensis,* which lives on several of the West Indian islands, lays exceptionally large eggs supplied with sufficient yolk for complete development to take place within the egg, and hatching occurs only when metamorphosis has been almost completed. Only the absorption of the tail takes place after emergence.

Perhaps the most unusual of the reproductive aberrations exhibited by frogs and toads is that of the paradox frog, *Pseudis paradoxis.* The adult is a rather unremarkable small frog between two and three inches in length, but attractive in appearance with an iridescent green skin. Its tadpole, however, is an astonishing creature. After it has hatched from the egg it grows and grows, and goes on growing until it approaches a foot in length. Indeed, very large specimens do not stop growing until they have achieved a length of fourteen inches. When these enormous tadpoles were first discovered a search was made for the super giant frogs which it was believed must develop from them, but they were never discovered. At their metamorphosis they shed their tails in sections at the rate of about one inch a day, instead of absorbing them as other frogs do. The final result of this process is the small adult *Pseudis*. The paradox frog is found in various ponds in Venezuela and in just one or two ponds in Trinidad.

With their complete lack of tail and the great development of their back legs for jumping and swimming, frogs and toads are highly specialised animals. The newts and salamanders, with their long bodies and well developed muscular tails, look much more plausible as animals intermediate between the fish and the reptiles. Externally, too, the changes that take place at metamorphosis are much less drastic than those that convert a frog or toad tadpole into an adult.

With the approach of winter newts go into hibernation, usually on land, where they hide away in holes in the ground, under stones or even under piles of leaves. When they wake up again towards the end of February they make for the nearest stretch of water in order to breed. Unlike frogs and toads they seem to have little or no homing instinct, being content with the first suitable stretch of water they come across. At this time the males are adorned with a temporary crest along the back of the upper surface of the tail, which functions as a kind of breeding dress; they also become more brightly coloured. The crest and the colouring seem to play an important part in the courtship displays which are an integral part of the preliminaries to mating. The courting male swims around the female, nibbling at various parts of her body and generally showing off, until she has been suitably aroused. These displays may or may not be followed by a short period of amplexus or mating embrace, but this does not lead to egg-laying and fertilisation as it does in frogs and toads.

At the climax of courtship the male sheds through his cloaca a mass of spermatozoa embedded in a mucous-like secretion produced by the lining of the cloaca. Spermatophores, as these parcels of spermatozoa are called, are produced by the males of a number of quite unrelated groups of animals. As soon as the male deposits his spermatophore the female swims over it and takes it into her cloaca. In view of the fact that she cannot possibly see what she is doing she exhibits remarkable accuracy in picking up the spermatophore. Once inside, the active spermatozoa proceed to wriggle out of the mucus and swim into a special cavity, the spermatheca or receptaculum seminis, where they remain until they are required to fertilise the eggs as they are laid, the empty spermatophore case then being shed.

There is a delay of between three and ten days after she has picked up the spermatophore before the female newt begins egg-laying. The eggs are passed down the oviduct one at a time, and each is fertilised by one of the spermatozoa before being laid. Each is then carefully wrapped for protection in the leaf of an aquatic plant, the edges of the chosen leaf being carefully folded by her hind feet while she deposits the egg in the crease so formed. The fold is maintained by the gelatinous covering of the egg, which sticks the two edges of the leaf together. Each female will lay between two and five hundred eggs at the rate of a dozen or so a day over several weeks.

The tadpole that finally hatches from the egg after a week or two is similar to that of a frog or toad, but is rather more elongated and has a longer tail. The branched external gills that begin to develop while the larva is still enclosed within the egg are not replaced by internal gills, as they are in the frog and toad tadpole, but are retained throughout larval life. Internal gills are not developed in the newt tadpole. The only visible changes that take place at metamorphosis are the absorption of the external gills, their function being taken over by the pair of lungs that become functional at this time, and the absorption of the tail fin. The muscular tail itself is of course retained throughout life. While the majority of newt tadpoles metamorphose in late summer and autumn, many of those that develop from eggs laid towards the end of the breeding season do not have time to complete their development during the same season, and spend the winter hibernating as tadpoles, finally metamorphosing the following spring.

Among the salamanders there is more reproductive variation than among the newts. The European salamanders, constituting the family Salamandrinae, have become more independent of water in their breeding habits than any other urodeles. The best known member of the family is the common spotted salamander, *Salamandra salamandra,* which is widely distributed throughout central and southern Europe. During the breeding season, which occurs in April and May, the spotted salamanders assemble for a short time in the nearest water. Their courtship is similar to that of the newts and there is the same production of spermatophores by the males, and collection by the females. This accomplished, they now leave the water, the fertilised eggs developing within the oviducts of the females instead of being laid. After about a week, when the eggs have hatched to small tadpoles, the females seek out some clear running water, at the edge of which they lie with only their hind parts submerged. Out pass the tadpoles into the water, each female producing about two dozen. At this stage the larvae already have well developed gills, and their limbs have begun to grow. Towards the end of the summer larval development is complete and metamorphosis occurs, after which the young salamanders now leave the water to begin their terrestrial existence.

The spotted salamander is found both on the plains and in the hills up to about two and a half thousand feet above sea level. In the Alps at higher altitudes, from three to ten thousand feet, its place is taken by the black Alpine salamander, *Salamandra atra.* In general habits it is similar to the spotted salamander, living concealed beneath stones and moss in woods and beside streams. In its reproduction, however, it has gone still further in its emancipation from the water. The only time throughout their lives that these salamanders enter the water is at the breeding season, for mating and exchanging the spermatophores, and the total time occupied by this is generally not more than a few hours.

Not only are the eggs retained within the oviduct until they hatch, but the tadpoles also spend the whole of their life there.

At first they develop gills, but these are eventually lost. When at last birth occurs it is not tadpoles but young salamanders that emerge, fully equipped to assume a terrestrial existence. Because their young do not have to face the hazards of an aquatic larval life, with the high rate of mortality which this usually entails, the Alpine salamanders do not need to produce so many offspring as other urodeles. Usually the female gives birth to only two young, one from each oviduct. To begin with, however, several dozen eggs are fertilised in each oviduct, but only one of these develops, the others providing it with food material as it grows.

At the other end of the scale are salamanders that spend the whole of their lives in water, becoming sexually mature and breeding while retaining larval structure. In these there is of course no metamorphosis. This phenomenon is known as neoteny, and is exhibited by two complete families, the Sirenidae and the Proteidae, as well as by certain members of some other families.

Among these the most interesting is the Mexican axolotl. Until 1865 this, too, was thought to be permanently neotenous. In 1864 six axolotls from Mexico, five males and one female, were acquired by the Jardin des Plantes in Paris. In the following February the female began laying eggs, which subsequently developed into tadpoles. During the following autumn, much to the surprise of those responsible for them, the adults all began to lose the gills which had always been regarded as a permanent feature of all axolotls. Within a few weeks they had achieved a complete metamorphosis, and came out of the water prepared for life as terrestrial salamanders.

These were so like the American tiger salamander, *Ambystoma tigrinum,* that there could be little doubt that the Mexican axolotl represented a permanently neotenous race of that species. More recently, however, it has been established that these historic Paris axolotls were in fact the neotenous larvae of a closely related species, *Siredon mexicanum,* which takes the place of the tiger salamander in Mexico.

CHAPTER FOUR

REPTILES

The zoological importance of the reptiles is twofold. They were the first vertebrates to solve completely the problems of terrestrial existence, and they gave rise in their turn to the two highest groups of vertebrates, birds and mammals. Without the intermediate reptilian stage it is difficult to see how these latter groups could have come into existence.

The problem of desiccation was only partially solved by the amphibians with their moist skins, and this, as we have seen, restricts their distribution to damp and shady places. The reptiles have completed the development of a waterproof skin, and can exist exposed even in the hottest tropical sunshine without any danger of losing excessive amounts of moisture. Their skin has also become covered with scales, which offer varying degrees of protection from attack from their enemies. In some types, notably the tortoises and crocodilians, these scales form a considerable armour plating.

A dry skin is of course of no use as an organ of respiration, so the reptiles have to rely entirely upon their lungs for breathing, and as a result the lungs are much better developed than the quite simple lungs of the amphibians. In their movement, too, the reptiles have at their disposal stronger and better developed limbs than the amphibians.

Perhaps the function in which the amphibians have become least adapted to life on land is their reproduction. With few exceptions, which we have already noted, they are completely dependent upon water at the breeding season. To develop a satisfactory method of breeding entirely on land from the typical aquatic method of breeding common to the fishes and

mphibians drastic modifications were necessary. The prob-
ems involved were quite considerable, the first being to
roduce an egg that would not dry up if laid out of water.
This could be achieved in two ways, each of which has been
dopted by different groups of higher vertebrates. One method
vas to cover the egg with some kind of waterproof shell, which
vould prevent water loss while it developed, while the other
vas to retain it within the oviduct until it hatched, in which
ase desiccation would be prevented by the moist oviduct
valls.

Both methods involved more elaborate breeding procedures.
nstead of the eggs being fertilised in the water after they and
he spermatozoa had been shed, a process necessitating only a
imple kind of mating to ensure that both products were
eleased at the same time and place, the new methods required
more intimate form of mating in which the males actually
nserted their spermatozoa into the female oviducts. The
nethod evolved by the reptiles was to cover the egg with a
hell, and this method was inherited from them by the birds.
The primitive mammals used the same method, but later this
vas modified, the shell being lost and the eggs retained within
he oviduct. Certain reptiles have anticipated this method and
ave become viviparous, retaining their eggs within the ovi-
luct. Their viviparity, however, is much less complex than
he highly elaborate method evolved by the mammals.

Today's living reptiles represent but a meagre remnant of
vhat was once the dominant group of animals. Taking the
xtinct and living reptiles together, six subclasses of the class
Reptilia are recognised, only three of which have living repre-
entatives. The oldest of these is the subclass Anapsida, which
ontained the 'stem reptiles' from which all the other reptilian
ubclasses were subsequently evolved. It is to this subclass that
he tortoises and turtles belong, giving them pride of place as
he oldest of the living reptiles. Together they are collectively
nown as the chelonians. The subclass Archosauria contained
ll the giant dinosaurs, and is represented today by the croco-
lilians. One archosaurian order gave rise to the birds. The

subclass Synapsida has no living representatives, but it is c
supreme importance because one of its orders, the Therapsida
finally gave rise to the ancestral mammals. The majority c
the living reptiles, the snakes, lizards and *Sphenodon*, belon
to the subclass Lepidosauria. The other two subclasses neithe
have living representatives nor gave rise by evolution to an
other kinds of animals. They are the Euryapsida, containin
the plesiosaurs, and the Ichthyopterygia or ichthyosaurs. Bot
groups were specialised for life in the seas, their limbs bein
modified as paddles.

The most important feature of chelonian reproduction is th
complete lack of concern on the part of the females of mos
species once they have laid their eggs. In terrestrial specie
the males usually have longer and thicker tails than th
females, and in the base of the tail the penis is hidden. Afte
fertilisation the females excavate nests in sand, earth or pile
of rotting vegetation, filling in the cavities after the eggs hav
been laid. In some species the females stand guard over th
nests, occasionally urinating over them. The possible reaso
for this behaviour is obscure. It may be to keep the eggs mois
or it may be to prevent the nest covering from becoming to
hard and so preventing the young chelonians from emergin
after they have hatched. The number of eggs laid varies ver
considerably from species to species. In some only a singl
egg is produced, whereas in others the clutch may consist c
as many as two hundred eggs.

The vast majority mate and lay eggs only once every yea
In some species at least it seems that once mating has take
place the spermatozoa can remain active within the oviduc
for a considerable time. In the American diamond-back tor
toise the female is capable of laying fertile eggs up to fou
years after her last mating.

The most spectacular chelonian reproduction occurs in th
marine turtles. At the breeding season turtles in vast number
forsake the open seas and approach certain limited sand
shore areas in tropical and subtropical coasts. At a favourabl

pportunity, usually on a bright, moonlit night, the females ome ashore in thousands, having previously mated offshore.

The first to give a full account of these breeding activities vas the great American naturalist James Audubon early in he nineteenth century.

> Once landed [Audubon wrote], each female turtle proceeds to form a hole in the sand, which she effects by removing it from under her body with her hind flippers, scooping it out with such dexterity that the sides seldom, if ever, fall in. The sand is raised alternately with each flipper, as with a large ladle, until it has accumulated behind her, when, supporting herself with her head and fore part on the ground fronting her body, she, with a spring from each flipper, sends the sand around, scattering it to the distance of several feet. In this manner the hole is dug to the depth of eighteen inches, or sometimes more than two feet. This labour I have seen performed in the short space of nine minutes. The eggs are then dropped one by one, and disposed in regular layers, to the number of one hundred and fifty, or sometimes nearly two hundred. The whole time spent in this part of the operation may be about twenty minutes. She now scrapes the loose sand back over the eggs, and so levels and smooths the surface that few persons on seeing the spot could imagine that anything had been done to it. This accomplished, she retreats to the water with all possible despatch, leaving the hatching of the eggs to the heat of the sand.

Each female usually lays three such clutches at intervals of wo or three weeks. When the young turtles hatch they work heir way up to the surface of the sand and then make straight or the sea. While still on shore large numbers of them fall victim to land crabs and sea birds, and when they reach the vater many more are caught by fish, so that despite the large umber of eggs laid, only a very few of each brood survive ong enough to become adult.

As with tortoises and turtles, all crocodilians lay eggs. There is a simple kind of courtship which takes place at night, when the males go about roaring and are very quarrelsome among themselves. Both sexes are provided with two pairs of glands whose secretions have a strong musky odour. One pair opens in the groin and the other on the lower jaws. They are most active at the breeding season, the scent being stronger in the males than in the females. It presumably acts as a stimulus to mating.

After mating the female lays her eggs in a nest, which may be a mound of leaves and other decaying vegetation, or a shallow depression dug in sand. In either case the eggs are covered over and left to incubate. In most species the female remains in the vicinity of the nest to guard it against other animals to whom the eggs would represent a desirable meal. Small mammals and large lizards are always on the lookout for a nest that has been left temporarily unguarded.

When the eggs finally hatch the young crocodilians make their way to the surface—unaided in most species, but there is evidence that females that lay their eggs in a sand nest, notably the Nile crocodile, do help their offspring at the time of hatching. These sand nests set very hard, unlike nests made of vegetation, and the newly hatched young would probably find it impossible to force their way out of them. As they hatch they are said to make croaking noises, which act as a signal for the mother to start digging them out. Once they have emerged, however, she loses all further interest in them, so that all young crocodilians have to fend for themselves almost from the moment of hatching.

In contrast to the crocodilians and the chelonians the snakes and lizards show considerable variety in their reproductive habits. Anatomically, the most unusual feature of the lizards and snakes is the possession by the male of two penes. These are usually referred to as hemipenes, from a mistaken early belief that they represented one penis divided longitudinally into two parts. They are however both complete. Also, unlike the penes of any other animals, they are located in the base of the tail rather than in the body, and when at rest are turned inside out like the finger of a glove that has been pushed in. Another peculiar feature is that, whereas the sperm duct normally passes through the centre of the penis, in snakes and lizards it merely forms a groove on its surface down which the spermatozoa travel to its extremity. During mating both penes are not brought into use at the same time, the one used depending upon which side of the male's body is brought into contact with the female.

Unlike snakes, as we shall see later, there are considerable differences between the sexes in lizards in the majority of species. In some lizards there are constant colour differences between the sexes, whereas in others these differences appear only at the breeding season, or are revealed only momentarily by some kind of display. But whatever method is used a lizard usually has no difficulty in distinguishing the sex of another lizard by sight.

The defence of territory during the breeding season is a well known phenomenon in birds, but it is also a feature of the reproductive season in most lizards, each male defending the territory occupied by himself and one or more females. The main biological purpose of this territorial instinct is to prevent overcrowding.

Modern authorities believe that the displays of males in the breeding season are designed not so much to excite the females as to warn off other males from their territory. If the invading male does not take the hint then a contest ensues, which is usually won by the male in possession.

In the initial stages of courtship the male nudges the female with his snout and may lick some part of her body. He then proceeds to bite her, and again any part of her body seems to suffice. Having stimulated her sufficiently he then seizes her by one side of her neck with his jaws, following this by arching his body over hers until one of his hemipenes lies close to her reproductive opening, into which it is then inserted. There is some evidence that once fertilisation has occurred the spermatozoa can remain alive in the female oviduct for at least several months, so enabling a number of batches of eggs to be fertilised following a single mating.

Among lizards some species lay eggs, whereas in others the fertilised eggs are retained within the oviduct while they develop. The distribution of these two methods seems to be quite haphazard, two quite closely related species often adopting different methods. The most outstanding example, though, is the so-called viviparous lizard *Lacerta vivipara,* which occurs right across Europe and central Asia. In some parts of its range it lays eggs, whereas in other areas the eggs develop in the oviduct, so that the young lizards are born alive.

In a few skinks true viviparity has been developed, in which the yolk sac of the developing egg forms an intimate contact with the oviduct wall, so that materials can be transferred directly from the maternal to the embryo circulation.

Although egg-laying lizards may guard their eggs while they develop, there is no evidence that they show any form of parental care once their offspring have been hatched or born.

Reproduction in snakes is generally similar to that in lizards, which is not really surprising considering how closely related they are. However, whereas in lizards the sexes usually differ in appearance, so that they can recognise each other easily by sight, this is not so with snakes. To distinguish one another they rely upon the sense of smell, but where the odour comes from is still not known for certain. Some experts believe it is given off from the skin, whereas others think that it is produced from special anal glands.

Parental care is unknown among snakes as it is among lizards, but incubation of eggs once they have been laid is much more common. The most important effect of incubation is to provide protection for the eggs, and in some species at least, notably the Indian and king cobras of Asia, both parents take it in turn to guard their eggs in this way. Among some species of pythons there is evidence that while the female is incubating her eggs the temperature of her body may be as much as six degrees fahrenheit above the surrounding air temperature, so that she is able to keep her eggs warmer than they would otherwise be.

Some snakes construct primitive nests for their eggs, though most of these are nothing more than suitable cavities in which the eggs can be dropped. Perhaps the favourite site for the construction of one of these nests is in leaf mould or other decaying vegetation, where natural decay produces a certain amount of heat for incubation. The European grass snake, *Natrix natrix*, is the best known example of a snake that recognises the value of such warmth. It is also noted for the fact that many females will gather at the same site and lay their eggs in a communal nest. Several hundred eggs are often found at the same site, and in exceptional cases as many as three thousand eggs have been found in a single dump, these representing the production of something like three hundred females.

The most advanced snake nest, however, is constructed by the king cobra, *Naja hannah*. From the outside it just looks like a heap of leaves, but within it is divided into two distinct compartments, the lower one containing the eggs, while the upper one houses the parent that is guarding them.

During the course of their development snake eggs absorb moisture and swell, so that the shell becomes subject to considerable tension, which probably helps the young snake to release itself when development is completed. The most important aid to escape, however, is the so-called egg tooth, a tooth-like structure that develops on the end of the snout of the embryo snake. As the fully developed young snake moves about within the egg this tooth slits the egg shell in many places, enabling it to effect its escape. Although they do not need it, the egg tooth also develops in the young of some viviparous snakes, suggesting that its evolution predates the adoption of viviparity. At the first moult, which may occur almost as soon as the young snake is hatched or born, but in any case within the first two weeks, the egg tooth is lost, and does not occur on the new skin.

CHAPTER FIVE

BIRDS

Although they are an extremely large group of animals, birds show a remarkable uniformity in their fundamental method of reproduction. From their reptilian ancestors they have inherited the large egg with its generous supply of food material in the form of yolk, which enables the young bird to reach a relatively advanced stage of development before it becomes necessary for it to break out of the shell, a process known as hatching. This shell, incidentally, has become hard and brittle by deposition of calcium carbonate in the soft leathery ancestral reptile shell.

Because of the relatively enormous size of the eggs compared with the size of the hen birds that lay them, there is not room in the body for the normal two ovaries characteristic of other vertebrates, so only one, usually the left, develops. The other is present only as a rudiment. If the functional ovary becomes destroyed by disease, or is removed by operation, the other ovary develops, but as a testis and not as an ovary. This sex reversal is also accompanied by a change in the appearance of the original female. Cock plumage develops as well as other male secondary sexual characteristics such as the comb.

Unlike reptiles, birds, like mammals, are homoiothermic, having evolved a method of keeping their body temperatures constant and fairly high, and this has an important bearing on the development of the eggs. In order to develop they must be kept at a similar temperature and not left, as in other animals, to develop at varying speeds as their temperature varies according to the temperature of their surroundings.

With the exception of one small group of Australasian birds

nown as the megapodes all birds keep their fertilised eggs at r near normal body temperature by sitting on them. And xcept for the majority of cuckoos it is the parents that under-ake this incubating chore. In certain species the female takes o further interest in her eggs once she has laid them, and eaves it entirely to the male to incubate them.

The nests in which the hens lay their eggs vary very much n complexity. Many species simply lay their eggs on the round, usually on gravel or stones. Some of them scrape out ome kind of hollow, whereas others make no preparation vhatsoever. But the majority do construct an actual nest vhich, while it may still be on the ground, is more likely to e in a shrub or a tree, or on an inaccessible cliff ledge. Almost very conceivable material is used, grass, leaves and other lant material, and mud, which may also be interwoven with lant material. One species of swift which lives in China and ther parts of the Far East constructs its nests entirely of aliva. These are the famous birds' nests from which soup is nade—a soup regarded as a great delicacy in these countries.

Many birds, notably owls and woodpeckers, take over holes n trees as their nesting sites. One of the most unusual hole esters is the hornbill, the bird whose enormous bill even utdoes that of the toucan. The female takes over a large tree avity in which she constructs a nest. Once she has laid her ggs the male bird proceeds to seal up the opening to the avity with mud, leaving only a small hole through which he an pass food to her and to her chicks after they have hatched. Only when they are fully fledged and ready to fly will the dult pair break down the mud wall to allow the hen bird and er chicks to emerge.

The grebes also show unusual nesting behaviour, for they build nests that float on the lakes and ponds on which they live. These nests are of course safe from the attentions of land-based predators, and they are in no danger of being over-whelmed by waves or swept away by water currents. Clearly, the grebes could not adopt this habit if they were river dwellers.

The number of eggs laid by different species varies con
siderably, and the economics of these egg numbers are ver
interesting. Large birds have a longer natural expectation o
life than small ones, a fact that is generally true for all animals
Also, large birds are less likely to be killed by predators than
small ones. For these two reasons large birds do not have to
produce so many offspring as small ones in order to keep thei
populations steady.

As against this, however, for very small birds there is an
optimum clutch size. Tiny nestlings lose heat very fast, and a
larger number of them huddled together in the nest will los
considerably less heat individually than a smaller number
and therefore each one will need less food. For example it ha
been shown that a small brood of three great tits require
three-quarters of the amount of food that will suffice for a
brood of eight. So with little more effort a pair of great tit
can bring up a brood of eight instead of one of three.

Some birds will also vary the number of eggs in a clutch
according to the availability of food. For example, short-eared
owls, which feed their young on voles, and snowy owls, which
feed theirs on lemmings, will lay more eggs in years when
their prey are abundant than they will in years when they are
scarce. If there is insufficient food to go round all the nestling
suffer and few will survive.

Although, as we have seen, reproduction in the majority o
birds fits into a general constant pattern, there are four group
whose reproductive behaviour shows significant variation from
this pattern. The remainder of the chapter is devoted to a
study of these aberrant types, which are the true flightles
birds, the penguins, the cuckoos and the brush turkeys and
their relatives.

The living members of the class Aves (birds) are considered
to represent three super-orders, the first of these being the
super-order Palaeognathae, a small group of true flightles
birds comprising the ostriches, emus, rheas, cassowaries and
the kiwis, and perhaps also including the hen-like tinamus o

South America. These birds, more familiarly known as the ratite birds, are considered on structural grounds to be the most primitive of all living birds, though the reasons for this judgement are beyond the scope of this book. It is not certain whether they represent the modern descendants of a group of birds that had never passed through a flying stage, or whether they diverged at an early stage of evolution from a primitive flying stock, although concensus of opinion favours the second view.

The kiwi is surely one of nature's most extraordinary creatures—a bird that not only cannot fly but can hardly see. Although it is the national emblem of its native New Zealand, most New Zealanders have never seen it, mainly because it is a completely nocturnal bird, living well away from human habitation in dense forests. During the daytime it hides in small caves or under fallen trees, coming out only at night to feed. In daylight it is practically blind, and there is some doubt as to whether it can see much better at night.

To make up for its lack of sight, however, it has a very strong bill which is extremely sensitive both to touch and to smell, the two senses on which the bird relies for nearly all its activities. When it comes out to feed it walks along continually tapping the ground with the tip of its bill, much as a blind man taps the pavement with his stick. It is searching for earthworms, which are almost the only food that it eats. As soon as it has detected one the bill is plunged into the ground and the worm is dragged out. So accurate is the bird's method of detection that it seldom has to make a second attempt.

The nostrils, unlike those of other birds, are placed at the very tip of the bill, where, of course, they are of most use. The efficiency of a kiwi's sense of smell has been demonstrated by putting an earthworm on the ground in front of it and letting the worm crawl away, holding the bird until the worm has travelled several yards. On being released the kiwi begins its characteristic tapping and sniffing, following the path taken by the worm with the certainty of a bloodhound until it has caught up with it.

The kiwi's breeding habits are no less interesting and unusual than its other activities. Nesting sites are similar to those chosen for sleeping, and the nest is very rough and ready. To begin with only a few twigs are collected, and perhaps a certain amount of excavating may be undertaken. In these preliminary preparations both parents co-operate. The female then lays one egg, which is incredibly large, weighing about a pound—something like a fifth of her own weight. Having achieved this mighty effort she loses all further interest and wanders off, leaving the male to incubate the egg, a long process taking about seventy-five days. During this time he rarely leaves the nest, and consequently gets very little to eat, so that by the time the egg hatches he has lost about a third of his original weight.

His duties are by no means over when the young chick at last appears, however. Although it is fully feathered it is not really strong enough to leave the nest to live on its own for a week or so. To keep it in the male barricades the entrance with sticks and leaves. When he does finally take the chick out he cannot of course see it if it strays far away from him, but he can always keep in touch with it by his sense of smell. Almost as soon as it is out of the nest the young bird starts digging for worms, its bill being already quite strong.

Until it is old enough to fend for itself the male guards the chick jealously, attacking any other bird that comes within reach. If it is frightened it always runs to its father for protection, but will have nothing to do with its mother, threatening her with its beak if she approaches.

Recent observations suggest that although the female kiwi is capable of producing only one of her monster eggs at a time, she may well produce another a month later. Exceptionally she may produce a third and even a fourth, also at monthly intervals. Such multiple laying is certainly tough on the male, because each succeeding egg laid after the first increases his incubation time by another month, so that it is possible for him to be engaged on incubating eggs and training young chicks for as long as six months at a time.

For chicks to be hatched fully feathered and active is unusual. Most chicks are completely helpless on hatching, and acquire their feathers and the ability to move about only after a further nestling period. With all the ratites, however, the prolonged development within the egg includes the nestling phase in addition to the normal incubation phase. It is probably for this reason that the eggs are abnormally large, enabling them to be provided with sufficient yolk for both phases.

Although they are only a small group, the flightless birds are widely distributed. The emus of Australia are second in size only to the African ostriches. Their nest is nothing more than a shallow hollow scooped out of the sandy soil, and in it about ten eggs are laid. Most if not all of the incubating falls to the lot of the male, and takes him something like sixty days. Even when they have hatched he is still responsible for looking after the chicks until they are well grown. Only then, when they are almost ready to fend for themselves, will the mother sometimes deign to relieve the male of his duties. In contrast to the rather drab grey–brown appearance of the adults the chicks with their striped plumage are very attractive little creatures, and make fascinating pets for a few months until they begin to grow up.

Despite the distance separating the two animals, the breeding habits of the South American rheas are remarkably similar to those of the Australian emus. The rheas do not, however, associate in pairs—each male bird gathers around him as many as seven females which he guards carefully, chasing off any other male who dares to approach. The females after mating lay their eggs in a communal nest which, like the emu's nest, is nothing more than a shallow depression scraped in the soil. Some females may lay their eggs outside the actual nest, in which case the male goes around rolling them into the nest area. Having assembled his complete clutch, which may number up to thirty or more, it is then up to him to incubate it without help from the females. The incubation period is about thirty days. After hatching the chicks remain with their father for some time until they have grown sufficiently to fend for themselves.

The breeding habits of the African ostrich, the largest of all the living ratites, are again similar in broad principles to those of the emus and rheas. Like the rhea each male ostrich usually mates with several females, who lay their eggs in a communal nest, a depression about a foot deep and three to four feet in diameter which is bulldozed out by the male using his breast. The number of eggs in the combined clutch may number as many as twenty. Incubation throughout the night is undertaken by the male, but the arrangements for daytime incubation depend upon the climate. In warmer parts incubation is unnecessary, the sun providing sufficient heat to keep the eggs warm, so they are merely covered with a layer of sand to protect them from predators. In cooler parts, however, where incubation has to be continued throughout the day, the females do give the male some relief.

Large though the existing ratites are compared with other birds, they are nevertheless only moderate in size when compared with some of their extinct ancestors—the moas, which once lived in New Zealand, and the largest of all the known fossil birds, *Aepyornis*, an inhabitant of Madagascar. This bird was probably the basis for the mythical roc featured in the adventures of Sindbad the Sailor. It certainly must have been a massive bird: fossil eggs of *Aepyornis* have a capacity of between two and three gallons, which makes them roughly equivalent in volume to six ostrich eggs; and the average ostrich egg weighs about four pounds! It would be interesting to know how long these giant eggs had to be incubated.

The second super-order of the class Aves, the Impennae, is another small group of specialised birds, the penguins. Early in their evolution they lost their power of flight and became highly specialised for life in water. Their wings became modified to form very efficient and powerful flippers, and their feet became webbed, though they play little part in swimming. They are confined to the southern hemisphere, and many of them live in or near the Antarctic.

At the breeding season all penguin species are gregarious, assembling in rookeries which may contain many thousands of breeding pairs. Those species that live within or close to the Antarctic Circle face the greatest problems, and the most important of these are the emperor, the king and the adélie penguins. Those living in warmer climates construct rough nests or burrows, and their general nesting habits are similar to those of other sea birds.

Apart from the problems of incubating their eggs in extremely low temperatures the Antarctic species also face the problem of hatching their eggs early enough for the young birds to become sufficiently well developed during the short summer season so that they can survive the following winter. The adélie penquins solve this problem by nesting early. They are small birds and prefer rocky slopes exposed to the wind where snow cannot accumulate and bury them. These slopes will be near to the coast. During the winter, however, vast areas of the Antarctic seas freeze over, so that in September and October the rookery sites will be many miles from the open sea. But this is the time when they must start nesting. And so they come out of the water and on to the ice, across which they trek to their rookeries, a journey of at least forty miles, and in some cases as much as two hundred miles.

The actual nest is nothing more than a rough pile of stones. At first a mound is raised by the birds lying on their stomachs and pushing backwards with their feet, and this is then added to by picking up pebbles in their beaks and placing them on top. One or two eggs are laid, and are incubated by the adults lying flat on them so that they are in direct contact with the brood patches.

Incubation is shared. Usually the male takes first shift while the female returns to the sea to feed. Depending upon how far the rookery is from the open sea she may be away for anything between seven and eighteen days. Thus, including the initial trek to the rookery and the time taken in nest-building and courtship, the male may well have fasted for as long as forty days before he is finally relieved and able to go off to feed. By

King penguin brooding its egg which is held on its feet under a fold of skin, the other in the act of tucking in the egg

this time of course the sea ice is melting, and he will not be away so long before returning to allow the female to go off once again. So the two continue to take turns throughout the thirty-six-day incubation period. When they first emerge from the sea they have considerable stores of fat laid down in the form of subcutaneous blubber, but after his forty days or so without food the male will have lost about one-third of his original weight.

Even after the chicks hatch the adults cannot let up because the young must now be supplied with an abundance of food so that they will be full grown and ready to moult before finally leaving their parents and going to sea for the first time in February.

In contrast to the adélie penguin, which has a length of about thirty inches and a weight of between nine and fourteen pounds during the breeding season, the emperor penguin, the largest of all the penguins, is about four feet in length and weighs anything between fifty and one hundred pounds. Because of its much greater size its eggs take longer to incubate and the chicks longer to mature. As a consequence emperors begin their breeding season four months before adélies.

During June, which is the middle of the long Antarctic winter of perpetual darkness, the emperors come out of the sea on to the sea ice. Prior to this they have been feeding incessantly so that they begin the breeding season with a layer of blubber or subcutaneous fat which may be as much as one and a half inches thick and which provides them with an abundant source of energy which they will need during the ensuing months. Their rookeries are always on the sea ice, but well away from areas where the ice meets the sea in the depths of winter. So, like the adélies, they face a long trek of at least sixty miles, and perhaps more, from the sea. By the end of June or early in July they will have reached the rookeries, and here they mate. As soon as the female has laid her single egg the male takes charge of it, because it falls to his lot to incubate it, while the female begins the long trek back to the sea to feed.

Emperors make no nest. It would be quite impossible to keep the egg warm if it were deposited on the ice, and there are no materials available that could be used to make a nest that might raise it above the ice. Instead the male penguin, standing in an upright position, balances the egg on the top of his feet, where it is covered from above by a fold of abdominal skin and held against the warm vascular blood patches of the abdomen. The efficiency of this arrangement is demonstrated by the fact that, while the surrounding temperature may be as low as minus forty degrees centigrade, the temperature inside the incubating eggs will be about thirty-one degrees centigrade, an astonishing difference of around seventy degrees centigrade.

Even so incubation takes a full six weeks, during which the

male remains virtually immobile, and of course can take no food because none is available. Even a few seconds' exposure to the air would cool the incubating egg sufficiently to kill it. In order to reduce heat losses the incubating male emperors huddle together in groups of as many as six thousand individuals, those on the outside of the huddle periodically pushing towards the middle to be replaced by others so that none remain exposed on the outside for very long. It has been calculated that huddling reduces the amount of heat lost by the body by about one-sixth compared with what it would be if the bird remained exposed all the time—a considerable saving over a period of six weeks.

As the chicks are due to hatch, the females, once again well fed and with a replenished layer of blubber, return to the rookery to take on the young chicks as they hatch and to allow the now emaciated males to return to the sea to feed and replenish their reserves. The chicks remain under the abdominal fold and are fed on regurgitated food, the females having returned not only well fed themselves but with about seven pounds of fish in their crops. Eventually, when they have become covered with a thick layer of soft grey down, the chicks are able to leave the protection of the abdominal folds. By the beginning of the Antarctic summer, in October and November, the now full-grown chicks are able to leave their parents and make for the sea which, due to the gradual melting of the ice, is not so far away as it was when their parents first left it in mid-winter. By the time they reach the water they will have moulted, exchanging their grey down for their sleek adult plumage.

In contrast to the emperors, the king penguins make no attempt to hatch and rear their young in one season. Eggs are laid in summer between October or November and January. The breeding season is preceded by a nuptial moult, by which time the adults will have spent two or three weeks at sea fattening themselves up for the long fast ahead of them. Like the emperors, the female kings lay only a single egg which is incubated on the feet of the parents. Unlike the emperors, both parents share the incubation duties. The incubation period is slightly shorter—about fifty-four days. By the time the eggs have hatched at about the turn of the year there is insufficient time for them to develop to full size and undergo the first moult which will provide them with adult plumage before the onset of winter.

Thus it is that king penguin chicks live through their first winter covered with their characteristic plumage of soft grey down, only moulting and becoming adult during their second summer. Because of this protracted development of their young, king penguins are not able to breed every year, though some of them do manage to raise a chick in two out of every three years. These are the pairs that succeed in producing their eggs towards the end of October.

Almost all birds incubate their eggs by sitting on them, thus providing the necessary warmth from their own bodies. Only two groups have managed to evade this rather arduous responsibility, the cuckoos and the megapodes. And the cuckoos—or at least the majority of them—have achieved this by getting other birds to do the job for them.

Cuckoos are widespread throughout the world, but wherever they are found their habits are similar. Basically they are the parasites of the bird world, getting other birds to undertake the arduous duties of incubating their eggs for them.

Cuckoos are fascinating creatures because of their unusual habits, and also because there are still more mysteries to be solved than facts that are known about them. Altogether there are more than one hundred and fifty cuckoo species in the world, and the vast majority of them are migratory birds. Once she has mated, as everyone knows, the female cuckoo lays her single egg in the nest of another bird. But this is not just a random process. It is now clear that each female is committed to a chosen host species. There are those that lay their eggs in hedge sparrows' nests, and those that lay their eggs in the nests of warblers, and so on.

In most cases there is definite egg mimicry, the egg laid by the cuckoo bearing a remarkable resemblance to the egg of the host. The cuckoos adapted to laying their eggs in the nests of any particular species constitute a group known as a *clan* or *gente*. Whatever clan it belongs to a mated female cuckoo doesn't merely fly around hoping to come across a suitable host nest by chance. Instead it watches birds of the required host species, and in this way is guided to the nest.

Presumably a young cuckoo is predisposed to lay its eggs in nests of the same species in which it was brought up, but whether this really is so is not known. Also what happens when two cuckoos belonging to different clans mate is not known. Do their offspring lay their eggs in the nests of the female's predisposed host species or the male's, or do they have a free choice?

Let us now consider an individual female cuckoo conditioned to lay her egg in a reed warbler's nest. She will fly around until she finally locates a pair of reed warblers constructing a nest. She then watches them until they have completed their nest and the female has begun to lay. This is the signal for her to fly down to the nest while the parents are temporarily absent, laying her own eggs directly into the nest and at the same time flying off with one of the warbler's eggs. In due course she will probably eat this stolen egg.

In earlier times these facts were misinterpreted. The fact that female cuckoos were seen with eggs in their mouths gave rise to the belief that these were their own eggs which they had laid elsewhere and were carrying to their host's nest.

Although the cuckoo is usually a much larger bird than its host, and would therefore be expected to have a longer incubation period, the incubation period of the cuckoo egg is in fact very short—about twelve and a half days, so that it may be expected to hatch before the eggs of the host species that are sharing the nest with it. Within a short time of hatching the young cuckoo proceeds to push the host's eggs or its hatched nestlings over the rim of the nest. Once these are outside the nest their own parents cease to take any interest in them, and from now on they devote all their activities to feeding the usurper, which in consequence gets the food which

Megapode-scrub turkey, *Alectura lathami*. The mounds can be 12–15ft in diameter and 6ft high

by instinct is designed to feed four or more offspring. As a result it grows very quickly and is soon ready to leave the nest.

There is one more mystery surrounding the cuckoo. It does bear a remarkable resemblance to the sparrow hawk. Small birds traditionally mob sparrow hawks and other birds of prey, and they also mob cuckoos. It is tempting to regard this resemblance as an interesting example of mimicry. Unfortunately it is difficult to see what advantage the cuckoo gains by resembling a hawk. More likely the resemblance is just a coincidence.

Only one group of birds has managed to dispense with the necessity of sitting on its eggs to incubate them. These are the Megapodiidae, a small specialised family belonging to the large order of game birds. Their eggs of course do need heat for incubation, but this is provided in the majority of species, as it is in some reptiles, by rotting vegetation. Most of the megapodes build large mounds of leaves and other plant materials, and when these have begun to decay and produce heat the female birds lay their eggs in the middle of the heap. Because they make use of artificial incubators, the megapodes are sometimes referred to as incubator birds. The different species are variously known as brush turkeys, scrub fowl and mallee fowl, and they are found only in Australia and certain neighbouring Pacific islands, notably New Guinea.

The nesting mounds can be extremely large, and are prepared by the birds in early spring. John Gould, the eighteenth-century authority on the birds of Australia, has left a classic description of these nest-building activities.

> The heap employed for this purpose is collected by the birds during several weeks previous to the period of laying; it varies in size from two to many cart-loads, and in most instances is of a pyramidal form. The materials composing these mounds are accumulated by the bird grasping a quantity in its foot and throwing it backwards to a common centre, the surface of the ground for a considerable distance being so completely scratched over that scarcely a leaf or a blade of grass is left. The eggs are deposited in a circle at the distance of nine or twelve inches from each other, and buried more than an arm's depth with the large end upwards.

These megapode mounds are usually thrown up in areas of sandy soil, and after the pile of leaves has become thoroughly wetted so that they can begin to rot it is covered with a layer of sand, which will serve to keep in the heat. Now begins one of the most remarkable phenomena in the whole of natural history. The month is September, and the female megapodes are ready to lay their first eggs. But before they do so they must test the temperature within the mound, which will by now have reached a level of between ninety and ninety-six degrees fahrenheit. To do this they open up the mound and take up a beakful of sand, which apparently enables them to judge the temperature within. Their aim is to achieve an internal temperature of ninety-two degrees fahrenheit. If they find that the temperature is above this they will not replace so much sand on the mound, and may even leave it completely uncovered for an hour or two. If however the temperature has not reached the required level, usually because the weather is cold and windy, they will heap the sand higher than it was before.

Once the mound has reached a steady temperature of ninety-two degrees fahrenheit the females prepare to lay their eggs. The ritual begins with the male opening up the mound and once more testing the temperature. If he is satisfied the female takes his place in the cavity he has excavated and proceeds to lay one egg. Before leaving she may scrape a little sand over it, but it is left to the male to finish burying it and close up the mound.

The eggs are always laid blunt end uppermost. This ensures that the head of the chick faces upwards when it finally hatches after an incubation period lasting for seven weeks or more. Like the kiwi, the megapode lays relatively enormous eggs. The average hen weighs about three and a half pounds, yet her egg is only slightly less than half a pound in weight. Exceptionally she may lay a second egg two days after laying her first, though the interval between successive eggs may be as long as seventeen days. The females often go on laying for a considerable time, some laying more than thirty eggs in a single season, representing a total egg weight more than three times their own body weight, no mean achievement.

All this means of course that the mound is in use for many months, but throughout this period the birds are so skilled at regulating the internal temperature of the mound that it remains at almost exactly ninety-two degrees fahrenheit.

By January most of the leaves will have rotted, and so they are now providing little heat. But this is the warmest part of the summer, and the late eggs that are being incubated at this time rely mainly upon the heat of the sun to maintain their incubation temperature. To allow for this the birds have to modify their behaviour. In the morning they scrape most of the sand off the mound and spread it around. In the afternoon, when it has become thoroughly warmed by the sun, it is heaped back on to the nest to keep it warm through the night.

The reason for the excessive size of the eggs is that they contain sufficient food material for advanced development to take place within the egg, so that when the chicks finally hatch they have reached the same stage that most chicks have reached after their period as nestlings. As a consequence they receive no parental care, merely pushing their way upwards until they emerge from the mound fully able to care for themselves. Within a few hours they are able to fly, and within three months they are full grown.

The members of one megapode species found in New Guinea have evolved a different method of keeping their incubating eggs warm. They nest around the volcanic Matupit crater near Rabaul, and instead of constructing mounds they dig burrows in the soil, which is a mixture of black volcanic soil and pumice-stone dust, and is still kept warm by the interior heat of the volcano. The native population dig the eggs out from the burrows to eat, but it is forbidden to kill the birds themselves.

CHAPTER SIX

MAMMALS

The mammals share with the birds the distinction of being the only two groups of animals that are homoiothermic, maintaining a constant and fairly high body temperature. In their reproduction, however, they have solved the problems of keeping their embryos at this temperature in a completely different way. With the exception of the monotremes and the marsupials, they have developed a method by which their developing eggs are retained within a specially modified oviduct where they can be nourished and kept warm. To achieve this, however, took some time, and the earliest mammals still retained the reptile type egg in which a large food store was incorporated, sufficient to provide the needs of the developing embryo until it was ready to hatch.

They had, however, already developed two important mammalian characteristics, the body covering of hair, which is not found in any other vertebrate, and the production of milk, a nutrient fluid that is used to feed the young mammal after it had hatched. These first mammals are still represented today by the two Australian monotremes, the platypus and the echidna.

The next stage in the evolution of reproductive methods in the mammals is shown by the marsupials. After a short period of development in the uterus, sometimes no more than a few days, a very immature embryo is born and immediately transferred to a marsupium or pouch, where it can be fed on milk until it becomes well developed. Finally came the eutherian or placental mammals in which the embryo was retained in the uterus for the whole of its embryonic development, being nourished through a special connection between the embryo and the wall of the oviduct known as a placenta.

The living monotremes, the duck-billed platypus and the spiny ant-eater or echidna, represent a very early stage in the evolution of the mammals from the reptilian ancestors. The platypus, *Ornithorhynchus*, is well adapted to the aquatic life that it leads in Australian and Tasmanian rivers, where it is not uncommon. It is nocturnal, coming out at nights to feed on the invertebrate life of the river and retiring during the day into burrows which it excavates in the river banks.

Prior to mating the female platypus lines her burrow with wet leaves gathered in the main from the surface of the river. The wetness is important, because the eggs that she lays after mating are reptilian in character, having a soft leathery shell which can easily lose water, and the moisture in the leaves prevents their drying up. Before they are laid, however, they are retained within her primitive oviduct for a gestation period of about fourteen days. This of course is not comparable to the gestation period of a placental mammal, for there is no connection between the wall of the oviduct and the developing embryo through which food materials could be transferred to the embryo.

When the eggs are finally laid the female coils her body around them to maintain their temperature. This period of incubation occupies about another twelve days, at the end of which time the young platypuses hatch. Although the total development period is so short, the succeeding nesting period is correspondingly long. For the first eleven weeks or so the young platypus is completely helpless and without sight. Even at the end of this period when the eyes at last become functional it will remain in the nest for a further six weeks before leaving it for the first time. During these four months it is fed on milk produced from modified sweat glands in the abdominal wall. These are not united in any way to open on nipples, the milk merely being released on to the hairs, from which it is licked up by the young platypus.

During the time she is feeding her one or two young the female platypus eats enormous quantities of food. The weight of a female varies between one and a half and two pounds, but at this time she may well consume almost her own weight of food in a single night. Once the young platypus has been weaned it soon acquires a healthy adult-type appetite.

There are two species of echidna, the Australian *Tachyglossus*, and *Zaglossus* which lives in New Guinea. Apart from the fact that it is nocturnal the echidna's general habits are quite different from those of the platypus. It is completely terrestrial, and lives in dry rocky areas where there is an abundance of ants and termites which it licks up with a long protrusible tongue similar to that possessed by the true ant-eaters in other parts of the world. During the winter it hibernates for about four months, burying itself in the ground by means of its sharp claws. During this hibernation its body temperature falls to little above that of its surroundings.

Only a single egg is produced, usually in September, and it is not retained for a period within the oviduct like the platypus eggs but is transferred as soon as it is laid to a temporary abdominal pouch, where it soon hatches as a naked and helpless youngster. Here it is fed on milk produced by modified sweat glands opening at the bases of two tufts of hair within the pouch. After several weeks the young echidna develops a covering of hair, and is then removed from the pouch and deposited in the nest where it is looked after by its mother for another month or two until it has developed sufficiently to be able to leave the nest and fend for itself.

The monotreme egg still contains an abundance of food material in the form of yolk to supply the needs of the growing embryo. But as the mammals developed further they lost this valuable food supply, and their eggs became very small and virtually without yolk, so it was necessary for them to develop alternative means of nourishing their developing offspring. The marsupial or pouched mammals have solved this problem in one way, the placental mammals in another.

With the exception of the true opossums, which are found in Central and South America and in the southern states of North America, the marsupials are confined to Australia, Tasmania and New Guinea, with a few introduced Australian species now flourishing in New Zealand. The marsupial egg is virtually devoid of yolk, but the fertilised egg is nourished during its short time in the oviduct by a milky fluid secreted by cells in the oviduct wall and absorbed by the embryo. There may even be a primitive connection between the embryo and the oviduct wall, through which a certain amount of food material can be absorbed from the maternal blood supply, but this method of transfer is very inefficient and not to be compared with the true placenta developed in the placental mammals.

After a minimum gestation period, which in the small opossums may last for no more than eight days, the extremely undeveloped and helpless marsupial babies are born. The stage of development they have reached is comparable with that reached by placental mammal embryos in the early stages of pregnancy. But in one respect their development has been rapid. The forelimbs and their claws are well developed compared with the rest of the body, and they now have an extremely important part to play in the next stage of their development.

Unlike the monotremes, the milk-producing glands of the marsupials, like those of the placental mammals, are concentrated to form a small number of pairs of true mammary glands, each opening on a teat, and these mammary glands with their teats are covered by an abdominal pouch or marsupium which gives the group its name. The precociously developed forelimbs and claws are used by the new-born marsupial to crawl up the abdomen of its mother and into her pouch. At birth the young opossum is smaller than a worker honey bee, and even in the largest kangaroos the new-born baby is not more than half an inch in length. It is said that the mother kangaroo, just before her babies are due to be born, licks her fur from the top of the pouch to the opening

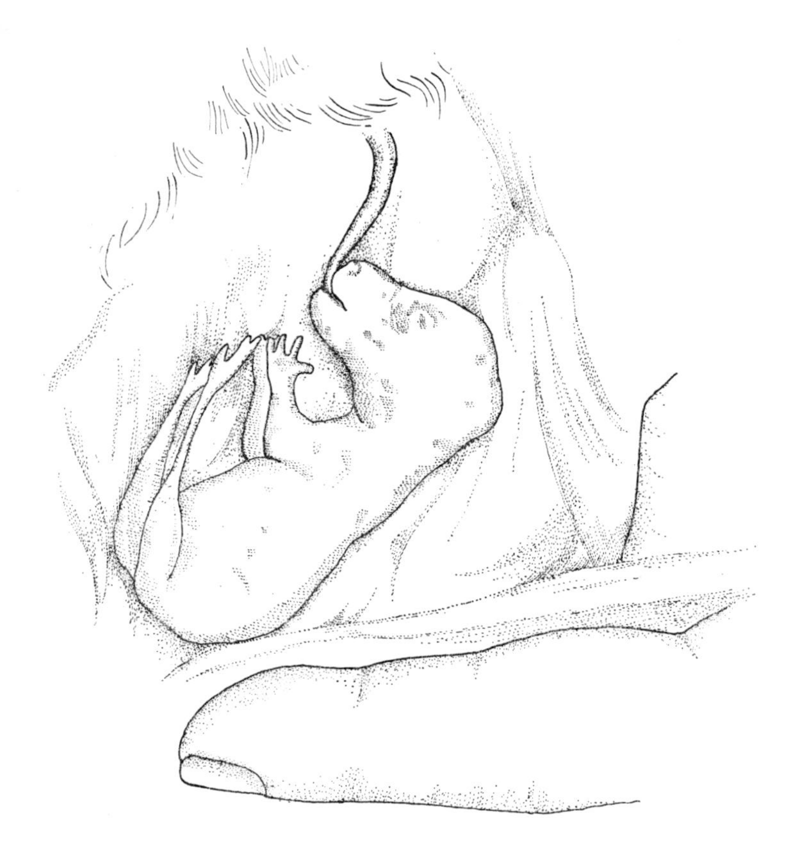

Kangaroo in pouch

of the oviduct, and that it is by following this saliva trail that the young kangaroos are able to find their way to the pouch.

Once in the pouch the young marsupials attach themselves each to one of the teats, and as the mouth closes around the teat the teat swells and, as it were, traps the mouth so that it cannot be withdrawn. And so for some weeks the youngsters remain permanently attached to their source of food, the teats pumping milk into them as fast as they can absorb it. When they are sufficiently developed, with fully formed limbs and a good covering of hair, they are finally released from their captivity, but for some considerable time they still remain in the pouch and go on feeding on milk. As they get bigger they will come out of the pouch from time to time and hop around, but if danger threatens they will dive back into it.

In the eutherian mammals development of the embryo takes place within the uterus, where it is nourished through a placenta, until the body of the young mammal is fully formed, when it is finally born. In many species it is still helpless, and often blind and hairless, though in a minority of species the young animal is able to run about within an hour or two of birth. All, however, have to be fed by their mothers on milk for some time before they are weaned and able to feed and otherwise fend for themselves. This period of post-birth dependence is also very variable, and in exceptional cases may extend for well over a year.

Because they are completely devoid of any food stores in the form of yolk, mammals' eggs are extremely small, usually less than one hundredth of an inch in diameter, and in consequence the ovaries that produce them are also small when compared with the ovaries of most other animals. The eggs or ova are contained within a sphere of cells, the whole constituting an ovarian follicle. When the egg is ripe, that is fully formed, and ready to be fertilised, the follicle absorbs quantities of fluid and so increases in size and bulges out on the surface of the ovary. The ovary itself opens into the funnel-shaped upper end of the oviduct, the fallopian tube, a narrow tube leading backwards into the uterus. The follicles finally burst, the egg with the contained fluid passing into the fallopian tube to travel down it to the uterus. It is during this passage through the fallopian tube that the eggs are fertilised by spermatozoa obtained by a mating at the appropriate time.

In the uterus the eggs begin developing. As the original single cell divides it forms a knot of cells at one end, which will form the eventual embryo, and a hollow sphere of cells surrounding it and a fluid-filled cavity. This is the blastocyst stage, and as we shall see later in certain species development is arrested at this stage, the blastocyst floating around in the uterus for a variable period before embarking upon the next phase of development.

While the follicles have been bursting to release their eggs and they have been travelling down the fallopian tube and becoming fertilised, changes have also been taking place in the wall of the uterus. This has become much thickened, and its blood supply considerably augmented, so that it now has an abundance of blood capillaries carrying a copious supply of blood. Except in those species in which development is temporarily arrested, the blastocysts now bury themselves in this enriched uterus wall, from the capillaries of which oxygen and the food materials needed for further development are obtained by diffusion.

Soon a more elaborate method of transfer of these materials develops, a placenta in which capillaries belonging to the maternal circulation are in intimate contact with the capillaries of the developing embryo circulation. The walls of these capillaries are so thin that the transfer of materials is easily achieved. It must be emphasised that the two circulations remain quite separate, no maternal blood entering the embryo blood vessels. As the embryo increases in size it bulges into the cavity of the uterus, and then becomes connected with the placenta by the umbilical cord, consisting of an umbilical artery and vein belonging to the embryo circulation.

The events leading to ovulation, the term used for the release of the eggs from the follicles, and the preparation of the uterus for the reception of the fertilised eggs are under the control of hormones, as is indeed the whole of the reproductive cycle. Most mammals do not breed all the year round, but have definite breeding seasons. Only at these times is the female mammal willing to mate, an occurrence known as oestrus or heat, the whole breeding cycle being the oestrous cycle.

As the egg cells mature the fluid within the follicles produces a hormone oestradiol which passes into the blood and causes the changes in behaviour which cause the female to permit mating. In some mammals the follicles burst spontaneously, whereas in others it only occurs following mating, and if no mating occurs the follicles remain intact and oestrus is prolonged until it does occur.

After bursting to release their eggs the follicles increase in size through an increase in the size of their cells and not through an increase in their numbers. Each follicle thus becomes transformed into a solid ball of cells known as a *corpus luteum*. Its function is to produce a second hormone, progesterone, which in conjunction with oestradiol is responsible for the changes in the wall of the uterus already described, thus ensuring that the uterus is ready to receive the fertilised eggs when they arrive after their passage down the fallopian tube.

Some mammals have only one oestrous cycle in each breeding season, whereas others may have several oestrous cycles in each season, mating again almost immediately after the birth of the previous litter. Species with only one cycle are said to be monoestrous, whereas those which have several cycles each season are described as polyoestrous.

In many male mammals there is a rutting season comparable with the female oestrus, when they become physiologically and psychologically attuned to reproductive activity. In any particular species of course rut and oestrus must coincide if mating is to take place.

The seasonal activity of the reproductive organs in the male and female mammals is controlled by hormones secreted by the pituitary gland, an extremely important endocrine organ situated at the base of the brain. In the female a follicle-stimulating hormone causes the ovarian follicles to develop, and a luteinising hormone stimulates the follicles to change to corporea lutea after they have burst to release their egg cells. Just prior to the birth of the young a third hormone, the lactogenic hormone, stimulates the mammary glands to produce milk. A hormone similar to the follicle-stimulating hormone is produced in the male, and this stimulates the testes to produce spermatozoa.

When we come to the ultimate question, what controls the

easonal production of these pituitary hormones, we find that here is no firm answer. It has been suggested that increasing mounts of daylight in the spring might stimulate pituitary ctivity, and this is almost certainly true in some cases. But ow could the amount of daylight affect nocturnal species, nd how could such a system work in the case of those mammals that breed in the autumn, when the amount of daylight s steadily decreasing? Much obviously remains to be learned bout the hormone control of the mammal's breeding cycle.

n the remainder of this chapter we shall consider one or two pecial aspects of mammalian reproduction.

In the majority of mammals the developing egg becomes mplanted in the wall of the uterus as soon as it has reached he blastocyst stage, and the embryo then continues to develop ntil it is ready to be born. The length of this gestation period s generally correlated with the size of the animal, small mammals reaching the stage when they are ready to be born more uickly than larger ones, though there are exceptions to this eneral rule. For instance, the rhino and the elephant take omething like two years to produce their weighty offspring ut the hippo, similar in its enormous size, completes its mbryonic development in the very short period of eight nonths, producing at the end of this time a well developed alf weighing not far short of one hundredweight. It isn't that lephants and rhinos are slow in their development, rather that the hippo is incredibly swift.

At the other end of the scale we have certain mammals whose development seems to be interminably prolonged. For instance why should a female bat or stoat take up to eight months to produce her minute offspring, or a female seal not many weeks short of a year? The explanation is not the same in all cases, nor is the reason for the delayed development always clear.

With seals both the explanation and the reason are easily provided. Seals are aquatic mammals, the adults being fully adapted to life in the sea. But unlike whales and dolphins they

have not become sufficiently readapted to an aquatic existenc
to be able to reproduce in water. Consequently they still hav
to come out of the water at the breeding season to mate, an
also for the females to produce their young. In order to avoi
two terrestrial excursions each year these two functions hav
become synchronised, so that as the pregnant females haul ou
on to the rocks to produce their offspring so the males als
make their appearance to mate with them almost immediatel
afterwards and so ensure the next year's brood.

But fifty weeks or so are not really needed for the develop
ment of seal pups. So the seals have adopted a system o
delayed implantation. When their eggs have reached th
blastocyst stage, by which time they will be about the size of
pin's head, instead of becoming embedded in the wall of th
uterus, as would happen with the majority of mammals, the
remain floating about in the cavity of the uterus for severa
months without undergoing any further development. Onl
after this dormant period do they finally become implante
in the wall of the uterus and resume normal development
which is completed just as the females haul out on to the rock
at the next breeding season.

Roe deer also make use of delayed implantation. Like fallow
and red deer they have a late summer or autumn rutting
season, but whereas the larger species require the whole of
the winter for the development of their offspring, the production of roe deer fawns can be completed in about five
months. Accordingly the fertilised eggs remain dormant at the
blastocyst stage from the August mating until December,
when they finally become embedded and resume development.
Again we can see the reason for delaying implantation. The
depths of the winter would hardly be conducive to reproductive behaviour, and if mating were delayed until early spring
the fawns would not be sufficiently well developed to be able
to face the following winter.

Delayed implantation occurs in the family Mustellidae,
which includes the stoats, weasels, martens and badgers, more
frequently than in any other mammal family. In the badger,

for instance, mating occurs between the middle of July and the end of August, but implantation is delayed at least until the end of December. The actual gestation period is only about eight weeks, and the young are usually born in March or April. Sometimes implantation is even further delayed. On one occasion a female badger captured and kept in a stable where she could not possibly have had any contact with a male produced twin cubs almost twelve months later. It has been suggested that this delay was induced by the shock of capture.

The situation in the stoat is particularly interesting. Young stoats are born between late March and early May, and within a few weeks the mothers will have mated again. Meantime their offspring will be growing rapidly, and by the time they are two months old, in June and July, they are mature and ready to mate. Implantation in both groups of females is then delayed until the following March, the embryos remaining dormant at the blastocyst stage for an incredibly long period of between seven and ten months. The actual gestation period is only about four weeks.

Why the stoat exhibits this peculiar reproductive timetable is a mystery. After all, most other small mammals, including the closely related weasel, get on perfectly well by mating early in the spring and producing their offspring at the same sort of time as the stoat. One curious feature of the stoat is that the male is fertile for only a few months of the year, from late March until July.

Experiments carried out with the American marten have thrown some light on the phenomenon of delayed implantation. Normally mating takes place during July and August and the young are born during the following April and May. But two American biologists, Pearson and Enders, have shown that if mated females are subjected to artificially increased light the total period of time between mating and birth can be shortened by between three and four months. Probably this increased illumination acts through the eye and the brain on the pituitary gland, which is responsible for secreting the hormones controlling development.

Delayed implantation is associated with a delay in the normal progressive development of the corpora lutea, the bodies which develop from the ovarian follicles as soon as these have released their eggs, and which are responsible for producing the hormones which control the development of the embryo. In those species exhibiting delayed implantation the corporea lutea remain in the immature stage of development that they normally reach just after the rupture of the follicles. Only after the appropriate delay do they begin to enlarge and produce the hormones that stimulate the wall of the uterus to prepare for the reception of the blastocysts. Presumably in the martens subjected to increased illumination this results in the renewed activity of the corporea lutea.

Prolonged development is not always achieved by delayed implantation. Bats mate in late summer before going into hibernation, but their single offspring are not produced until the following spring. In their case, however, it is fertilisation and not implantation which is delayed. After mating the sperms are stored right through the winter in the uterus. Only in the spring are the eggs released from the ovaries to be fertilised by these stored sperms. The gestation period which follows occupies about six weeks.

In some species of bats events are even more curious. The sperms that were acquired by the previous autumn mating are expelled before the eggs are released from the ovary. A second spring mating now occurs, and it is the sperms received at this mating that are used to fertilise the eggs. These sperms have also been stored right through the winter, but in the body of the male. The reason for both types of sperm storage is that the bat testis is only functional and active in late summer and autumn, and is quite incapable of producing sperms in the spring when they are really required, a curious state of affairs for which no reason is known.

Some mammals at the breeding season show highly organised social behaviour, and none more so than the various species of deer. They are also of interest in that, instead of having their breeding season in the spring, as many mammals lo, it occurs in the autumn. Another remarkable feature is the nnual growth and shedding of antlers. In many mammal pecies the males develop secondary sexual characteristics vhich distinguish them from the females, and this antler rowth represents this phenomenon in an extreme form.

Social behaviour among deer has been studied most inten- ively in the red deer (*Cervus elaphus*) of the Scottish high- ands, but the events are similar in other species. A herd of ed deer is a definite social unit. Except at the autumn rutting eason, which lasts for about six weeks, the sexes remain eparate. The hind (female) herds are usually fairly large roups, which contain all the young stags up to two or three ears old as well as hinds of all ages. These hind herds are isually led by one of their number, who is always very alert o detect danger while the others are feeding, and who leads he herd away if danger threatens. The members of these hind ierds seem to be held together by strong social ties,

Stags on the other hand form much less definite associations, hough they do tend to remain in groups, which are usually nuch smaller than the hind herds. Each hind herd usually ccupies a definite territory, which may extend over several quare miles. Stags occupy different territories from the hinds. The difference between the two groups is clearly seen when hey are disturbed. A hind herd moves off as a compact body, ll the members taking the same direction following the leader, whereas a group of stags will often scatter in all directions.

It is at the rutting season that the social behaviour of all deer reaches its greatest complexity. In red deer the rut begins towards the end of September with significant changes in the behaviour of the stags. Through the remainder of the year they have not only tolerated one another but have associated peacefully together. Now they become intolerant and restless. The groups break up and their members go off one by one to the hind territories in search of the hind herds, the largest stags setting off first, as they are usually the first to come into rut. When a stag finds a group of hinds that have not been

claimed by another stag he takes possession. From now unti
the end of the rut the stags will feed very little, spending thei
time herding the hinds and roaring. Stags seem to be silen
except during the rutting season, whereas the hinds are cap
able of uttering a kind of bark when alarmed at any time o
the year. The stag's roar is in the nature of a warning to othe
stags who may be within ear-shot to keep away from his harem

When the rutting season begins the velvet on the stag'
antlers has already been lost. They now consist of hard solic
bone, and can be used if need be to chase off intruding stag
or even to indulge in some fighting. This aspect of the rutting
season has often been much exaggerated. Serious fighting doe
not often occur, and battles in which a stag is killed are
exceptional. A preliminary skirmish is usually sufficient to
convince one stag of the other's superior strength, and he
retires before any serious fighting breaks out.

In this way it happens that at the beginning of the rutting
season the strongest stags each become possessed of a substan-
tial harem, which by continual chivvying they manage to keep
within a limited territory. The weaker stags patrol the areas
around these territories, roaring frequent challenges but
usually exercising sufficient discretion to refrain from following them up.

A stag in possession of a hind herd is constantly active. As
the hinds graze quietly he is for ever running around them,
preventing any from straying away from the group, rather like
a sheepdog with a flock of sheep. All the time he is giving vent
to an occasional roar, and is always on the alert for any other
stags in the neighbourhood which might approach his harem
too closely. He rarely lies down, even for a short time.

Although during the rutting season the hinds appear quite
indifferent to the stag that has temporarily taken charge of
them, they do nevertheless play an important part in the social
pattern of the rut. Deer hinds come into oestrus at different
times, and until they have done so they will not permit mating. A few will already be in this state at the beginning of the
rut, and they will be served first. The herding by the stag, and

even the roaring of the other stags in the neighbourhood and their encounters with their own stag, probably all act as a stimulus to other hinds, causing them to come into oestrus.

The first stags to gain possession of harems cannot maintain sexual activity for the full six weeks of the rutting season. After three or four weeks they are exhausted and retire, to be replaced by other stags. Even before this those stags which started off with very large harems will have gradually lost small groups of hinds to other stags.

The stag taking over in the latter part of the rutting season is able to serve those hinds that come into oestrus late. Old hinds tend to develop extremely late oestrus, and are sometimes not in heat until December or even January. These will be served by young stags, who would stand no chance of acquiring hinds of their own during the proper rutting season.

Many hoofed animals grow permanent horns, which are primarily used as weapons of defence or attack against other species. The antlers grown by stags are remarkable in that they are produced solely for combat with other stags during the rutting season, and are not normally used as weapons against enemies. When the rut is over they are shed, and new antlers are grown the following year. These become ready for use only just before the rut.

The growth of these antlers is an astonishing process. All that remains after the old antlers have been shed are two knobs of bone, which are part of the top of the skull. They are covered by the skin of the head. It is on these knobs or pedicles that the antlers are formed. The first sign of the growth of new antlers is the formation of two velvety knobs on these pedicles. In some species these develop within a fortnight of the shedding of the old antlers. Their velvety appearance is due to the fact that the skin with which they are covered carries a thick pile of fine hairs. The skin itself is richly supplied with blood vessels which makes it very sensitive and tender, and warm to the touch. It has the remarkable property of being able to form large quantities of bone. The antlers are laid down as a solid core of bone within the velvet.

Once these knobs have appeared, the growth of the antler is rapid, and in about three months development is complete Some idea of the rate of bone formation can be gained fron the fact that a really big pair of red deer antlers may consist o seventy pounds or more of solid bone. As the antlers grow, s does the covering skin, so that when growth is completed th whole antler is still covered with velvet. Because it is so sensi tive stags are careful to avoid knocking their antlers agains anything while they are in velvet.

As soon as the antlers are fully grown a ring of bone is lai down around the base of each, just above the point where i joins the pedicle. This burr, as it is called, presses outward and stretches the overlying velvet, gradually constricting th blood vessel passing out to the antler, and finally stopping th flow of blood altogether. Deprived of its blood supply, th velvet soon dies and becomes dried up. It is of course no longe warm or sensitive. The stag now proceeds to rub its antler against trees and other suitable objects until the dead velve is all stripped off.

Whatever the physiological mechanism controlling antle development may be, it is obviously closely bound up with th mechanism controlling reproductive activity, because the ru always follows soon after the antlers are out of velvet. Hov soon after the rut the antlers are shed depends upon th species, and shows great variation even between differen individuals of the same species.

The mechanism of antler shedding is as interesting as tha of their growth. The base of the antler, between the burr an the pedicle, is gradually eaten away by special cells in th blood. When the process is complete the antler, including th burr, just drops off, leaving only the pedicle.

Two groups of mammals have become remarkably wel adapted to life in the sea: the whales, dolphins and porpoises which comprise the order Cetacea, and the seals, sea-lions an walruses, which form the suborder Pinnepedia, one of th three orders into which the order Carnivora is divided. Th whales are completely adapted to an aquatic existence, whereas the seals and their relatives have to pay brief visits to land in order to breed.

Whales mate in the water, and the growth of their embryos is rapid. The largest of all whales is the blue whale, full-grown specimens of which will approach one hundred feet in length. The gestation period of this species is similar to that for most other whales, about twelve months. At birth the young whale will already have attained a length of something like twenty-five feet and a weight of between twelve and fifteen tons, easily the fastest growth rate to be found anywhere in the animal or plant kingdoms. Within two years of birth it will have grown to about seventy feet in length and be approaching sexual maturity.

One might imagine that suckling under water would be difficult, but in fact it is achieved quite simply. The female has a pair of teats concealed in grooves far back on the underside of the abdomen on either side of the vagina. When the calf wants to suckle the teats are protruded and it takes one into its mouth. The mammary gland is surrounded by muscle, which now contracts to force milk into the calf's mouth. The milk cannot take the wrong turning and get into the lungs because in the whales the larynx is inserted into the hind end of the nostrils at the back of the throat. This also prevents water from passing down into the lungs when the whale is feeding.

Seals and their relatives, unlike whales, have not been able to solve the problem of breeding in the sea. In order to make sure that males and females do come together at the breeding season most species repair to a limited number of breeding grounds where the adults haul out of the water, often in immense numbers which may in a few species exceed a million individuals. As already mentioned, by means of delayed implantation the birth of the calves conceived during the previous year's mating season are not born until the females have returned to the breeding grounds to mate again. This of course obviates the necessity for two excursions on to land, one for mating and one for giving birth to the young pups.

Compared with most other mammals the development of seal pups is extremely rapid, an adaptation that means that they are soon weaned and able to take to the sea to lead independent lives. And this in turn means that the females need only spend a minimum time on shore, mating within a short time of giving birth.

Before coming ashore seals have been feeding hard and have accumulated a thick layer of blubber, the name given to the layer of fat that all aquatic mammals develop beneath the skin. This serves both as an insulation against the cold water and a valuable store of food. Seal pups are born in a quite advanced state of development. Many of them have already shed their milk teeth and replaced them with their adult set before they are born, and some have even shed their puppy coat while still in the uterus. But they are extremely thin, having virtually no layer of blubber, and those that have no thick woolly puppy coat can often be seen shivering violently for the first day or two.

Seal pups, though, grow at an incredibly fast rate, the blubber of the mother being converted into an extremely rich milk which is taken up by the pups as fast as they can absorb and utilise it. In some species the milk contains more than fifty per cent of fat and a high concentration of proteins, but no carbohydrates. This enables the pup not only to grow fast but to accumulate its own thick layer of blubber, much of the milk fat being laid straight down as a fat store for future use. In a short three weeks the pups of many species have trebled their birth weight, and have been weaned. For the next six to nine months there may be little further increase in weight.

CHAPTER SEVEN

INTRODUCING THE INVERTEBRATES

The members of the animal kingdom are divided into about twenty main groups or phyla, the exact number depending upon which authority you follow, because there is still considerable controversy among the experts about the classification of the smaller and more obscure groups. Concerning the main phyla there is virtually no difference of opinion, and it is only these dominant phyla that we are concerned with in this book.

In the first six chapters we have been concerned with the members of only one phylum, the Chordata, an extremely important group because it contains the vast majority of all the larger animals, and also perhaps because it is the phylum to which we ourselves belong. But in terms of numbers of species, it contains only about five per cent of all known animal species.

The phylum Chordata can be defined precisely, because all the members have one feature in common, the possession of a notochord. But when we turn to the rest of the animal kingdom, universally referred to as the Invertebrata, we find no common denominator. The members of this group are linked only by a negative characteristic—they do not possess a notochord. The very name Invertebrata is, strictly speaking, a misnomer: the lower chordates are also invertebrate, since they, too, lack a vertebral column. Invertebrata, however, carries the sanction of centuries, and could not now be replaced by the more accurate Inchordata or non-chordates.

Within the Chordata there is a fairly clear evolutionary succession so that, for example, the structural and physiological changes that converted an ancestral fish into the first undoubted amphibian or an ancestral amphibian into the first definite reptile can be pretty accurately described, even though the intermediate stages have long since become extinct. But within the Invertebrata the hypothetical links connecting the various phyla are, to say the least, nebulous. Indeed, they are classified as phyla because the gaps between them are so much greater than those between the various chordate classes.

But there are certain evolutionary advances that are pretty clear, so that we can recognise, as it were, certain evolutionary levels. Thus the phylum Coelenterata, with a body plan based upon two fundamental cell layers, must have preceded the phylum Platyhelminthes, in whose bodies a third fundamental cell layer appears sandwiched between the original two layers. This does not necessarily mean, however, that the coelenterates were the ancestors of the phylum Platyhelminthes. All we can say is that the Platyhelminthes show a more advanced state of organisation than the Coelenterata. Both, indeed, may have had a common ancestor with two cell layers, one group of common ancestors having given rise to the coelenterates and another to the flatworms.

If we now postulate that the common ancestor, having given rise to two more efficient groups, succumbs to their competition and becomes extinct, a most likely course of events, we are left with nothing tangible to link these two groups. And if we further imagine that similar evolutionary events took place over a period of at least five hundred million years, leaving us each time only with the more efficient products of common ancestors, we have some idea of the disconnected nature of our existing invertebrate phyla.

With the vertebrates the links that have become extinct as living animals have nevertheless left fossils enabling us to piece together the whole story. But the majority of invertebrates had no hard parts that could have been preserved as fossils to help us piece together the evolutionary story.

Hence it is that when we study the living invertebrate phyla we are presented with a number of separate groups whose relationships we can only guess at, helped sometimes by striking similarities in the methods of reproduction adopted by two different phyla, giving additional circumstantial evidence as to their affinities.

CHAPTER EIGHT

SPONGES, SEA ANEMONES AND JELLY-FISH

The sponges, which constitute the phylum Porifera, are the most primitive and least highly organised of all the multicellular animals. None of them has solved the problems of living on land, and the vast majority are marine, only a small number of species being found in fresh water. A few have managed to adapt themselves to life on the sea shore, but even these only occur near the low tide mark. They are thus exposed to the air only for a short time each day, and even then they are shielded from direct sunlight because they are found only under rock ledges or in the shade of overhanging rocks.

Most sponges are rather indeterminate in shape, many of them forming incrustations spreading over the rock upon which they are growing. Although some of them are capable of moving their positions on the rock slightly, they are essentially sedentary. They are in fact so unlike typical animals that it was not until the early part of the nineteenth century that it was finally established that they did indeed belong to the animal kingdom.

In his *Year at the Shore,* published in 1865, the great Victorian naturalist Philip Henry Gosse writes of the sponges as 'the most debatable forms of life, long denied a right to stand in the animal ranks at all, and even still admitted there doubtfully and grudgingly by some excellent naturalists. Yet such they certainly are, established beyond reasonable controversy as true and proper examples of animal life.'

The final proof that sponges really were animals, deriving

nourishment from organic matter and not making their food from simple inorganic substances as plants do, was one of the classic examples of nineteenth-century natural history investigation. In 1825 Dr Robert Grant of Edinburgh embarked upon the work that was to link his name inseparably with the sponges. This is how he described what he saw when he watched a living sponge under the microscope:

> In the month of November I put a small branch of the *Spongia coalita,* with a little sea-water, into a watch-glass, and placed it under a microscope; when, on reflecting the light of a candle up through the fluid, I soon perceived that there was some movement going on among the opaque particles floating through the water. On moving the watch-glass, so as to bring one of the apertures on the side of the sponge fully into view, I beheld, for the first time, the splendid spectacle it presented, a living fountain vomiting forth from a circular cavity an impetuous torrent of fluid, and hurling along in rapid succession opaque masses, which it strewed everywhere around. The beauty and novelty of such a scene in the animal kingdom long arrested my attention, but after twenty-five minutes of constant observation I was obliged to withdraw my eyes, from fatigue, without having seen the torrent for one instant change its direction, or diminish in the slightest degree the rapidity of its course. I continued to watch the same orifice, at short intervals, for five hours, sometimes observing it for a quarter of an hour at a time, but still the stream rolled on with a constant and equable velocity. About the end of this time, however, I observed the current become perceptibly more languid, the opaque flocculi which were thrown out with such impetuosity at the beginning were now propelled to a shorter distance from the orifice, and fell to the bottom of the fluid within the sphere of vision, and in one hour more the current had entirely ceased.

This water current first recorded by Grant is of supreme

importance to a sponge, providing it with everything it requires, as well as being the channel through which it gets rid of all unwanted substances produced within it. The sponge body, in fact, consists of little more than a series of canals through which this vital current of water flows. Like so many other sedentary aquatic animals the sponges are suspension feeders, living on the microscopic plankton animals and plants that are suspended in the water, and are removed from it during its passage through their bodies.

All over the surface of a sponge there are numerous tiny openings called ostia or pores through which the water enters. The canals into which these ostia open lead in turn into wider, flagellated chambers lined with special cells, called collared cells or choanocytes. Each choanocyte bears a single whip-like thread, called a flagellum, which is surrounded by a short collar and projects out into the water. The continuous beating of these flagella in the flagellated chambers draws in the steady current of water through the ostia. The collared cells are also responsible for trapping and absorbing the plankton organisms, and so providing the sponge with its food. After the trapped organisms have been digested within these cells, and the resulting products distributed to the other cells of the sponge body, any indigestible remains are passed back into the chamber to be carried out of the body by the water current as it leaves.

Each flagellated chamber has many canals opening into it, but itself opens to the surface by a single opening, the osculum or vent, much larger than the ostia, and often situated at the top of a prominence, and so looking rather like the mouth of a volcano.

Although apparently so complex, with its system of canals and special cells, the sponge is really very primitive in its organisation. It has no nervous system to co-ordinate its activities, and although the collared cells cause the continuous water movements through its body, they work quite independently of each other, and not rhythmically as do the similar ciliated cells in other groups of animals. There are no sense organs and no special provisions for respiration, each cell obtaining the oxygen it requires by simple diffusion from the water as it flows past.

In many respects the sponge body behaves rather like a group of single-celled animals living together as a colony than as a true multicellular animal. In this connection it is interesting that the collared cells are virtually identical with the members of the protozoan family Choanoflagellata, the collar flagellates. Quite often these form colonies of individuals.

Although so primitive, sponges show a remarkable variety in their methods of reproduction, both sexual and asexual. In sexual reproduction the eggs or ova may be fertilised either while they are still within the parent sponge body, or in the water after they have passed out through the oscula. In the former case the spermatozoa presumably are brought into the sponge body with the inhalant current through the ostia. In neither case are there any special reproductive organs.

In both the viviparous and the oviparous species the fertilised eggs develop into oval embryos, each enclosed within a capsule formed from a single layer of cells. Within this capsule the outer embryo cells develop flagella whose lashing movements cause the embryo to rotate and eventually burst out of the capsule. Only then in the case of the viviparous species do these free-swimming embryos become released into the exhalant canals to be carried to the exterior by the exhalant current.

For the next twenty-four hours or so they swim actively with a spiral cork-screw like motion, but at the end of this time the flagella are withdrawn and the larvae sink to the sea bed. If they settle on rock or stone they will be able to continue their development, but the vast majority which fall on sand, mud, seaweed or other unsuitable surfaces will perish. The successful minority undergo a metamorphosis which converts them into semi-transparent platelets of tissue known as postlarvae.

Although apparently static, these postlarvae are capable of moving very slowly over the rock surface by a flowing movement similar to that by which *Amoeba* moves and which for this reason is referred to as amoeboid movement. During these movements two individuals may meet and coalesce, later perhaps tearing themselves apart in such a way that one larva flows off carrying with it part of the body of the other. Eventually each larva will develop a single osculum, a sign that it has settled at last in its permanent position and that the post-larval phase is over.

With all higher animals there is not the slightest difficulty in deciding what constitutes an individual, but with sponges it is nothing like so easy. If many young sponges develop close together on the same rock surface they may well fuse together completely as they grow and spread over the rock, forming in this way an apparently single sponge derived from many fertilised eggs. Thus with sponges it is not really possible to speak of an individual, only of a sponge body in the sense of a sponge that looks like an individual. It is possible to produce such a single sponge body by placing a number of adult sponges close to one another in an aquarium, when they will coalesce completely as they grow.

Further evidence of the lack of individuality in sponges can be gained by pressing a living sponge through fine silk. The cells become separated from one another, the whole mass coming through the silk as a milky fluid. Examined under a microscope this fluid is seen to consist of single cells and small groups of cells, none of them organised into any sort of pattern. Left in the water, however, these separated cells gradually come together in groups or regenerative masses, each group eventually organising itself into a complete sponge body. In the experimental aquarium these regenerative cell masses have been known after a number of months to become sexually mature and produce viable larvae.

Asexual reproduction occurs in a variety of different ways. The most straightforward of these is fragmentation, by which the parent sponge body merely divides into a number of fragments, each of these then developing into a complete new sponge body. Slightly more complicated is reproduction by the method of budding, in which buds grow out from the parent body, separate from it and develop as separate individuals.

Freshwater sponges have a special method of asexual reproduction in which reproductive bodies called gemmules are produced. Gemmule production begins with the assembly of considerable numbers of amoebocytes—cells capable of moving about the sponge body with an amoeboid motion—each charged with granules that are believed to be stores of food. These food-rich amoebocytes organise themselves into spherical masses around which other amoebocytes lay down horny protective capsules. The gemmules remain dormant within the parent body until this eventually dies off as conditions become progressively more adverse with the approach of winter. Until the spring they remain within the dead parent body, when each comes into active life, flows out of its capsule and searches for a suitable site on which to fix itself and grow into an adult sponge.

The most remarkable of all the asexual methods of reproduction found in the sponges is undoubtedly the formation of asexual larvae. These originate like gemmules by the assembly of numbers of amoebocytes to form spherical masses, which then proceed to develop in exactly the same way as sexually produced larvae, forming flagella with which they can swim about until they settle as semi-transparent plate-like post-larvae.

That multicellular animals were originally derived by evolution from unicellular phylum Protozoa is acknowledged. From the close resemblance between the collar cells of the sponges and the collar flagellates it seems very likely that the sponges were derived from them. It is also believed, however, that all other multicellular animals were evolved from some other group or groups of Protozoa. Thus it would seem that the sponges did not in their turn give rise to any more complex types of animals. They represent, as it were, an evolutionary side-line, an experiment that was not proceeded with. For this reason, and assuming that the remainder of the animal kingdom resulted from a steady progressive evolution from its protozoan ancestors, the sponges are sometimes regarded as forming a separate subkingdom of the whole animal kingdom.

By contrast with the Porifera, the members of the phylum Coelenterata show much better organisation, though they are still comparatively primitive animals. They do behave as individuals, however, and their activities, both internal and external, are controlled by a simple nervous system, so that no group of cells is independent of the rest of the body.

It does not follow that because an animal is primitive in structure and occupies a place near the bottom of the evolutionary ladder it must of necessity be a relatively unsuccessful animal. The coelenterates are in fact an extremely successful phylum, occurring in the seas of the world in great numbers and in a variety of types. A small number have become adapted to life in fresh water.

Perhaps the most interesting feature of this type is that any particular coelenterate can be referred to one or other of two quite distinct forms. One of these is a sedentary animal with a cylindrical body, the best known being the sea anemones, while the other is a free-living animal drifting in the water and having a disc or a bell shape, and typified by the various kinds of jelly-fish. The sedentary cylindrical type is known as a polyp, and the free-living disc is called a medusa.

The body of all coelenterates consists of just two layers of cells, an outer ectoderm and an inner endoderm separated by a layer of jelly called the mesogloea. In the polyp the mesogloea is quite thin. The polyp body is hollow, the endoderm surrounding the enteron or gastric cavity which functions as a primitive stomach. The base of the polyp is attached to the substratum, while the opposite end forms a cone-like structure, the oral cone, in the centre of which is the mouth opening into the enteron. Surrounding the oral cone is a ring of tentacles, whose numbers vary considerably between different species. With their formidable batteries of sting cells, these tentacles are responsible for capturing the coelenterate's food and conveying it through the mouth into the enteron, where it will be digested through the activities of the endoderm cells. There is no second opening to the enteron, all indigestible remains being passed back through the mouth at the completion of digestion.

The relationship between the sedentary polyps and the free-swimming medusae can best be understood if we imagine a polyp becoming detached from the substratum, turning upside down and floating up into the water. Its body is then flattened until it becomes a disc-shaped umbrella, and the mesogloea greatly thickened until it constitutes almost all the bulk of the body. The tentacles may remain as a ring around the mouth, or may have migrated to form a fringe around the edge of the umbrella. In some species both rings of tentacles are represented.

Although sea anemones and jelly-fish are the best known of the coelenterates they are both very specialised. The best member of the phylum to study to get a true picture of the relationship between the two types of organisation, the polyp and the medusa, and an understanding of the fundamentals of reproduction within the phylum is *Obelia.*

A piece of seaweed detached from beyond the low tide mark and cast up on the shore will often be found to have growing on it what appears to be a miniature forest of strange looking plants an inch or two in height. Looked at under a powerful lens or a microscope each stem is seen to be a branched colony of tiny animals, each growing at the end of a short side branch of the main stem. Most of these are typical polyps, each having a ring of tentacles around the mouth. They are the feeding polyps for the whole colony. An opening in the base of the polyp connects with the general cavity that runs right through the main stem and its branches, through which the products of digestion of any polyp can be carried away to be used by polyps in any other part of the colony. As with the solitary polyp, however, the indigestible remains are passed out through the mouth of the polyp that captured and digested the food organism.

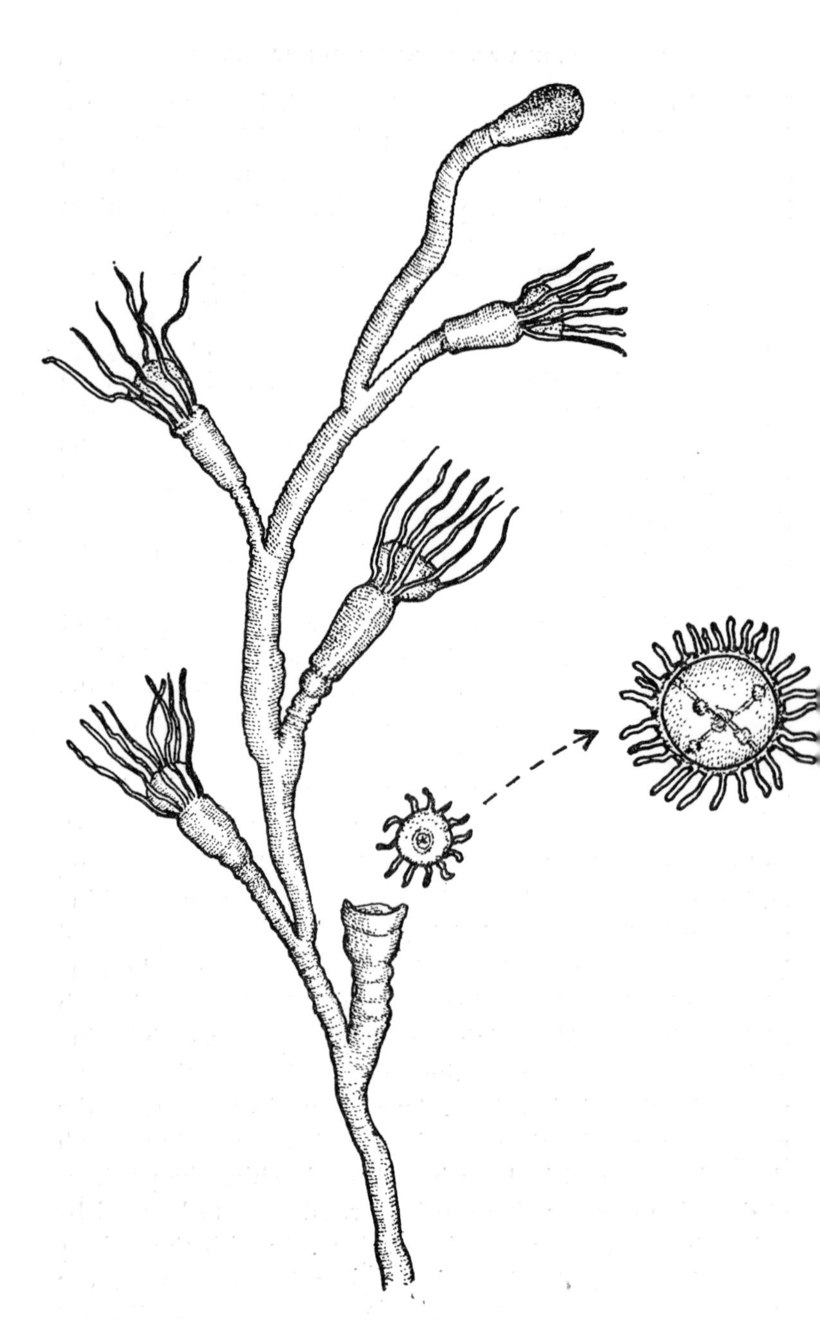

Obelia polyp with blastostyle releasing medusa, and medusa in later stage

Surrounding both the stem and the polyps is a transparent layer of horny material produced by the cells of the ectoderm and known as the coenosarc. This serves both to protect and strengthen the colony. Around the polyps it takes the form of a cup open at the top to allow the tentacles to project into the water.

Obelia can reproduce by three distinct methods, two asexual and one sexual. The simplest of these is asexual reproduction by budding, by which buds grow out from the main stem to form additional polyps and thus increase the total number of individuals in the colony.

The second asexual method is much more complicated. Some of the side branches in every colony develop as special club-shaped reproductive polyps instead of as typical feeding polyps. These blastostyles, as they are called, lack both mouth and tentacles, and rely for their food supplies upon the activities of the feeding polyps. On the surface of each blastostyle there are a number of saucer-shaped buds growing on short stalks. These are at various stages of development, those near the top being large and more complex in structure than those nearer the base. If their development is followed it will be seen that each is in fact destined to form a tiny medusa complete with a ring of tentacles around the edge of the saucer-shaped umbrella, and as each reaches the completion of its development it becomes detached from the central core of the blastostyle and swims out through the open top of the coenosarc cup and into the surrounding sea.

This medusa, produced from the *Obelia* colony by a special asexual method, itself represents the sexual phase of the species. The mouth opens at the end of a tube which hangs down from the middle of the under-surface and leads upwards into the enteron, which occupies the central portion of the medusa body. From it four canals radiate towards the edge of the saucer. Beneath each of these four canals a reproductive organ, formed from the ectoderm, hangs down into the water —four testes producing spermatozoa if the medusa is male, four ovaries producing ova or eggs if it is female.

There is no mating. The eggs and the spermatozoa are simply shed into the sea, where the eggs will be fertilised by the actively-swimming spermatozoa. Each fertilised egg now develops to form a kind of larva which is typical of the coelenterates and is known as a planula larva. It is oval in shape with a solid core of endoderm cells surrounded by a single layer of ectoderm cells. These are each provided with numerous cilia, tiny hair-like projections of protoplasm capable of beating rhythmically and so propelling the larva through the water. After a time it ceases to swim and sinks to the sea bed. If it comes to rest on a rock or on seaweed it can continue its development, but the vast majority that land on sand or mud will perish.

The settled planula soon reorganises itself and grows into a single polyp, the ectoderm cells withdrawing their cilia. First a cavity appears in the mass of endoderm cells to form the enteron, which becomes connected with the sea through a mouth which perforates the upper end, and around which a ring of tentacles develops. The single polyp then gives rise to a complete colony by the elongation of its lower end to form the main stem. This stem then proceeds to develop other polyps by budding, a process that has already been described.

The complete life history of *Obelia* illustrates two interesting phenomena that are well represented in the coelenterates, polymorphism and alternation of generations. Polymorphism literally means 'many forms', and is used to describe the occurence of two or more quite distinct types of individual animal existing within the same species. It is a phenomenon by no means confined to the coelenterates, but is an important feature of the phylum. *Obelia* possesses individuals of three separate types, the feeding polyp, the blastostyle and the medusa. Alternation of generations is the term used to describe the condition in which the members of one generation reproduce asexually to produce a second generation which produces sexually to give rise once more to the generation that produces asexually. In the case of *Obelia* the polyp represents one generation and the medusa the other. Again this is a phenomenon occurring in other phyla besides the coelenterates. The medusa generation plays an important role in dispersal. The sedentary *Obelia* colonies are unable to move to occupy any other suitable rock or seaweed surfaces on which they might be able to flourish. But the medusae are able to travel considerable distances from the parent colony, so that the planula larvae produced by them can colonise new areas.

The phylum Coelenterata contains three classes. The class Hydrozoa contains the members that show both polyps and medusae well developed. *Obelia* belongs to the order Calyptoblastea or Leptomedusae. The members of the order Gymnoblastea or Anthomeduseae are very similar, except that the coenosarc does not invest the polyps, stopping short at their bases, and the medusa is bell- or bud-shaped instead of being saucer-shaped.

The third order, the Hydrida, is extremely interesting from two points of view. Its most important members are the various species of *Hydra*, the best known of all coelenterates to students of biology throughout the world, and the most important of the coelenterates that live in fresh water. *Hydra* is a solitary polyp which neither buds to produce a colony nor produces a medusa. The medusoid phase in fact has been suppressed, and is now represented by testes and ovaries which are merely reproductive buds on the side of the parent animal. The ovaries each produce one egg which is fertilised *in situ* by spermatozoa produced by testes from another individual. *Hydra* is in fact hermaphrodite, that is each individual bears both male and female reproductive organs, but self fertilisation is avoided because the two kinds of reproductive organs mature at different times.

Hydra does in fact reproduce asexually by budding, but as soon as they are fully formed the buds become detached from the parent body to form a new solitary individual. Like so many of the lower animals, *Hydra* is capable of regeneration, by which a part of the body separated from the rest can develop to form a complete new individual. Experimentally it has been found that if an individual *Hydra* is cut into a number of pieces each of these can regenerate a complete new individual provided only that each piece contains at least one representative of each of the types of ectoderm and endoderm cells present in the complete animal. This was discovered more than two centuries ago by the Abbé Trembley, who was the first man to study *Hydra* intensively. He it was in fact who gave it its name, after the mythological Greek monster Hydra: this had many heads, and if one was cut off two grew in its place.

Trembley also made another claim which later authorities dismissed as improbable fantasy. He maintained that he had succeeded in turning a *Hydra* inside out, and that it continued to live a normal life. It was not until this century that his claim was finally vindicated, when an American biologist tried the same experiment, and his specimen also continued to live. He was able to discover the reason. All the cells of the ectoderm and the endoderm proceeded to migrate, so that in a short time the original ectoderm cells were once again on the outside, and the original endoderm cells were on the inside.

The second class, the Scyphozoa or Scyphomedusae, comprises the jelly-fish, in which the medusoid phase is dominant. The best known member of the class is *Aurelia,* a jelly-fish that occurs in almost all the seas of the world. It is quite a small jelly-fish, the average full grown specimen measuring about twelve inches across, with exceptional individuals attaining a diameter of two feet. The reproductive organs consist of four horse-shoe shaped bodies situated on the floor of the enteron and derived from the endoderm. The sexes are separate. In the males the spermatozoa are shed into the enteron and passed out through the mouth, to be taken in through the mouths of the females to fertilise the eggs in their reproductive organs. The fertilised eggs develop to form ciliated planula larvae while still within the female body, but are then shed into the sea. For further development they depend upon settling upon a rock surface, where they change to form a polyp-like structure.

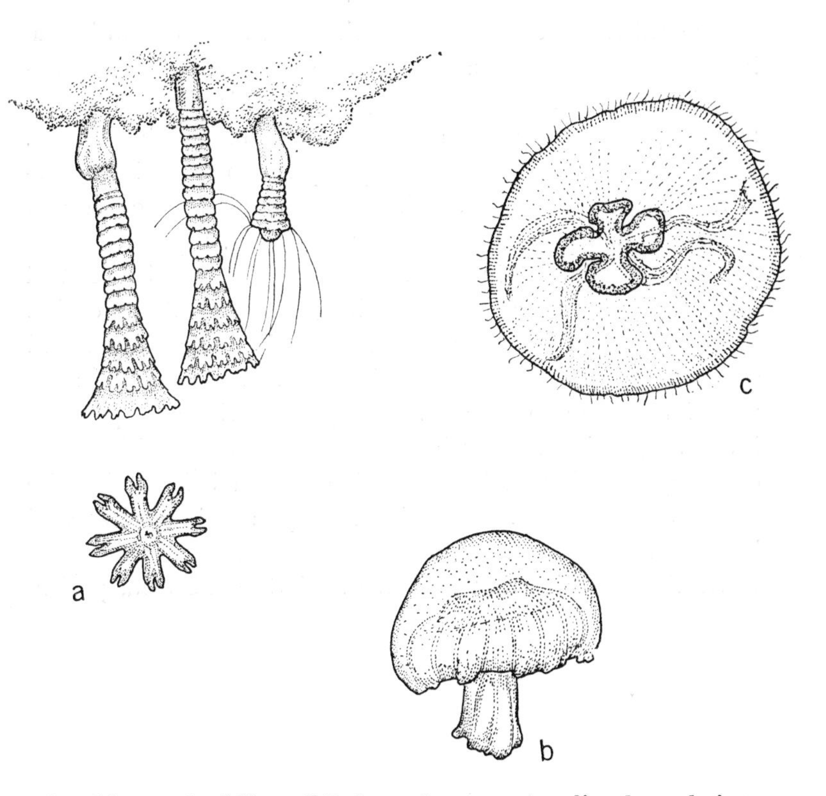

a Scyphistoma budding off Ephyra; *b* young *Aurelia*—lateral view; *c* adult *Aurelia*

At first these polyps possess a ring of tentacles like any other normal polyp, but these are soon lost. The upper part of the body then becomes constricted by a number of horizontal rings so that it looks rather like a pile of saucers. Each of these saucers then develops eight tentacles, while the constrictions grow deeper. Finally these little medusae become completely cut off one by one, and swim off into the sea to grow into normal adult jelly-fish. This polyp-like phase in the life history of a jelly-fish is known as a scyphistoma, and the process by which it buds off medusae is known as strobilation. In a minority of the Scyphozoa the fertilised eggs develop directly to medusae without the intermediate scyphistoma phase.

The third coelenterate class, the Actinozoa or Anthozoa, includes sea anemones and corals. In contrast to the Scyphozoa, in which the medusoid generation is dominant and the polyp generation much reduced or even altogether absent, the Actinozoa consist entirely of polyps, the medusoid generation being entirely suppressed. The enteron is a more complex structure than it is in the other two classes, its cavity being partially subdivided into a number of chambers by vertical folds of tissue that almost meet in the middle of the cavity.

Reproduction in the sea anemones is quite simple and straightforward. The reproductive organs are formed on the partitions of the enteron, and the reproductive cells pass out into the sea via the mouth. Here the eggs are fertilised and develop into planula larvae, which after a brief active life settle on the sea bed. Those that fall upon rocks survive, and become transformed into the next generation of sea anemones, while the majority that fall on unsuitable substrata perish.

CHAPTER NINE

FLATWORMS AND ROUNDWORMS

If you collect some of the vegetation growing in a pond, canal or stream and transfer it to a glass tank of water you will almost certainly find that within an hour or two the bottom and sides are liberally sprinkled with a number of elongated, flattened creatures gliding over the surface of the glass with no immediately apparent means of locomotion. Their length will vary from one-quarter of an inch to an inch, and their colour from white through various shades of brown to black. These are planarians, free-living members of the phylum Platyhelminthes or flatworms. An alternative method of collecting them is to put a piece of meat into almost any stretch of fresh water, withdraw it an hour or two later and similarly transfer it to an observation tank.

Although the flatworms are not to be regarded as highly developed members of the animal kingdom, they do represent one very important structural advance over the coelenterates which we met in the previous chapter. The flatworm body consists of three distinct cell layers, an ectoderm and endoderm, corresponding to the similarly-named outer and inner layers of the coelenterate, and a third layer of cells between them called the mesoderm, replacing the structureless mesogloea.

The phylum Platyhelminthes comprises three distinct classes, the members of only one of which, the Turbellaria, are free-living. The freshwater planarians are among the best known members of this class. All the members of the other two classes, the Trematoda, or flukes, and the Cestoda, or tapeworms, are parasites.

The turbellarian body is covered with cilia, and it is their rhythmical beating that carries it along smoothly without any muscular movement. The planarians live in fresh water but many of the other turbellarians are marine.

The usual position for the mouth of an animal is at the head end, but in some of the turbellarians, including the planarians, the mouth has migrated to the middle of the under-surface. Food is absorbed through a tubular pharynx which can be protruded when required through the mouth. Planarian food is entirely animal, and can be detected from some considerable distance by means of special sensory cells situated in the head region. As soon as contact is made with suitable food the pharynx is protruded and proceeds to tear off pieces from the victim by a powerful sucking action, which carries them into the planarian's digestive system. The mouth is the only opening from the digestive system to the exterior, so that any undigested remains also have to be passed outwards through it, as in the coelenterates.

All flatworms are hermaphrodite, and in contrast to the coelenterates their reproductive organs are extremely complicated. The planarian has a pair of ovaries, one on each side of the body just behind the head region, and from each an oviduct or egg duct runs almost the whole length of the body, numerous yolk glands opening into it at intervals all along its length. Eventually both oviducts open into a single genital chamber lying in the mid-line towards the hind end of the body. Also along each side of the body is a whole series of testes, those of each side opening into a single sperm duct or vas deferens. These both lead into a single muscular penis which also opens into the genital chamber. Also leading out from the genital chamber is a blind sac, the copulatory sac, and the chamber itself opens to the exterior by the genital pore situated on the under-surface of the body towards the hind end.

Although the planarian is hermaphrodite self fertilisation does not occur. To copulate two worms come together so that their under-surfaces are in contact. The penis of each is then protruded through its own genital pore and into that of its partner in such a way that it comes to rest in the entrance to the copulatory sac, where it deposits its spermatozoa. When mating has been completed the penes are withdrawn and the two individuals separate.

The spermatozoa then leave the copulatory sac and travel up the oviduct until they reach the ovary, where they fertilise the ripe eggs. These are then released to pass down the oviduct, and at the same time yolk cells, charged with concentrated stores of food, are also passed from the yolk glands into the oviduct. When the eggs and yolk cells reach the genital chamber they are arranged in groups, each group being surrounded by a shell to form an egg capsule. Each capsule contains usually less than ten eggs, but thousands of yolk cells. Yolk to provide food for the developing embryo is usually stored in the egg itself, but in the planarians the yolk is contained in these special yolk cells, the eggs themselves being devoid of yolk. The egg capsules when they pass out through the genital pore are attached to submerged stones and plants. There is no larval stage, the eggs developing directly into tiny replicas of the adults that produced them, emerging from the capsules to lead a free existence and fend for themselves when the food stores in the yolk cells have been used up.

Some planarian species are able to reproduce asexually by a simple division of the body into two parts. The first sign that a planarian is about to divide in this way is the appearance of a constriction of the body somewhere behind the level of the pharynx. This constriction intensifies until it has become an extremely thin waist. The hind end then seems to fix itself to the substratum while the head end tries to move on, resulting in the waist being severed. Both pieces then proceed to regenerate the parts that are missing, so that before long two complete individuals have been produced from the original single body. In certain species in which asexual reproduction is common sexual reproduction seldom occurs; some of them in fact produce sexual reproductive organs only rarely.

Because of their readiness to regenerate, the planarians

have been the subject of much fundamental research into the phenomenon of regeneration. In all cases it is found that a piece of a worm shows definite polarity, a new head always developing from the cut end that originally faced the front end of the animal, the new tail likewise growing from the cut end that originally faced the hind end.

Regeneration experiments also show that the capacity for regeneration is greatest near the front end, and decreases towards the tail end. In some species a section cut from near the tail end is incapable of forming a new head, and even if it does so in other species the growth is much slower and usually less complete. If the front end of a planarian is cut down the middle, thus dividing the head into two parts that are connected only further back, and are prevented from growing together again, two distinct complete heads are formed. Using this method it has been possible to produce a freak planarian with no fewer than ten complete heads all joined to the same original body!

The members of the second class of flatworms, the Trematoda, are all parasitic in their habits, and they fall into two groups: external parasites, which live on the surface of their hosts, and internal parasites, which live within their bodies. The shape of the trematode body is similar to that of the turbellarian body, and the gut also has but a single opening to the exterior. The body surface lacks cilia, but is provided with one or more suckers with which it can attach itself to its host Although they are also nearly all hermaphrodite, many trematodes do indulge in self fertilisation.

The classic example of the trematode is the liver fluke, *Fasciola hepatica.* Like most of the internal trematode parasites it has a very complicated life history, involving an intermediate as well as a final host, and produces enormous numbers of eggs. The production of large numbers of offspring is necessary for all internal parasites because of the remote chance of any one of them being able to find a new host in which to live.

Adult liver flukes live in the liver and bile ducts of sheep, goats and cattle and cause the often-fatal disease of liver rot. The liver fluke body is extremely flattened and leaf-like, about one inch in length and half an inch in width. There are two suckers with which it maintains its position, and a mouth through which it sucks the blood of its host.

Each adult fluke produces enormous numbers of eggs which pass down the bile duct into the intestine and so reach the exterior with the faeces. Those that are deposited in or beside a pond hatch out to form a ciliated active miracidium larva, which swims around in search of the intermediate host. In Britain this is the dwarf pond snail, *Limnaea truncatula*, but *Fasciola* is of almost world-wide distribution, and in different countries it has different intermediate hosts; in the United States, for example, it will use various species of *Galba*. The vast majority of the eggs, which will not of course be deposited near water, will perish, as indeed will many of those that are deposited in water unless they can find a suitable intermediate host within a few hours.

The successful miracidium makes its way into the snail's lung, where it changes into a sporocyst, a hollow sac in which numbers of a second kind of larva, called a redia, develop. These are soon released from the sporocyst and make their way into the body of the snail, most of them ending up in its liver. What happens now depends upon the time of year. In the autumn each redia eventually gives rise to a second generation of rediae, and these may even give rise to a third generation during the winter. During the summer, however, each redia produces a number of a third type of larva, known as a cercaria. These look rather like minute tadpoles, with tails by means of which they can swim. They leave the host body and swim around in search of a blade of grass. Those that successfully climb to the top of a blade lose their tails and surround their bodies with a protective covering. For many this is the end of the road, but those that happen to be eaten by sheep lose their protective covering when they reach their host's stomach and make for its liver, where they develop into adult flukes. Some idea of the rate of reproduction in *Fasciola* can be gained from the fact that a single miracidium larva can eventually give rise to up to six hundred cercariae. The second and third generation rediae which develop within the body of the intermediate host during the winter also eventually produce cercariae the following spring.

In tropical countries there is an important type of fluke which is parasitic in man and causes the disease known as bilharzia or schistosomiasis, which is prevalent both in Africa and Asia. Several species of *Schistosomum* are responsible in different areas. *Schistosomum* is unusual among flukes because the sexes are separate, though once he has found his mate the male carries her around with him permanently trapped in a fold of his ventral body wall.

Schistosomum lives in the blood vessels of the intestine, where it lays its eggs. As a blood vessel becomes distended with the number of eggs within it it ruptures, and the eggs are thus discharged into the cavity of the intestine. From there they leave the body with the faeces. With a modern sewage system *Schistosomum* would cause little trouble. But in many parts of the East, including China and Japan, human faeces are used as fertiliser for the soil, sewage being collected into reservoirs beside canals and irrigation ditches into which it is admitted as required. The eggs thus have ready access to water, where they hatch to form miracidium larvae. Within twenty-four hours they must find a water snail or perish. Those that succeed bore their way into the snail to feed on its tissues, at the same time losing their cilia to become transformed into sporocysts, which in their turn reproduce to form tailed cercariae. These make their way out of the snail and swim about near the surface of the water. If they make contact with the leg of a man or woman, which they are most likely to do in the paddy fields during the planting of rice, they attach themselves to the skin and bore their way into a blood vessel. It is now quite easy for them to travel round in the blood stream until they reach one of the veins of the intestinal region, where the life cycle begins all over again.

In its early stages schistosomiasis causes body pains, a rash and a cough, but later it becomes responsible for severe dysentery and anaemia. A victim may live for years, but will become gradually weaker and more emaciated, dying either of exhaustion or of some disease contracted through weakness.

The members of the third class of flatworms, the Cestoda or tapeworms, are perhaps even more highly adapted, especially in structure, to a parasitic mode of life than the Trematoda. Two tapeworms are most likely to be found in the human intestine, the so-called pork tapeworm, *Taenia solium,* and the beef tapeworm, *Taenia saginata,* pigs and cattle being the respective intermediate hosts for the larval stages. In earlier times the pork tapeworm was more common, which explains why most elementary textbooks of biology deal with it. Today, however, it has become almost extinct in America and Britain, leaving *Taenia saginata* as the more common tapeworm parasite of man in these countries.

The front end of *Taenia saginata* consists of a minute head or scolex provided with four adhesive suckers, by means of which the tapeworm anchors itself to the intestinal walls and prevents itself being swept away. In some species, including *Taenia solium,* there is also a circle of hooks to reinforce the activities of the suckers. Behind the scolex the body consists of a large number of segments or proglottides, which are being continually budded off from the hind end of the scolex. As they mature and become more separated from the head these proglottides increase in size. When they first become separated off from the neck of the scolex they lack reproductive organs, but as they grow bigger each proglottis develops a complete set of both male and female organs. These, too, are complicated, like those of the Trematoda.

After fertilisation of the eggs within the proglottis all the other structures within it degenerate, so that as it approaches the hind end of the tapeworm, which it does because those behind it are continually being shed, it becomes nothing more than a container full of mature eggs. Eventually in its turn it becomes detached from the parent body and so passes out with its host's faeces.

The eggs within the proglottis will remain dormant unless it is eaten by the intermediate host. In the host's stomach the protective shell with which each egg is provided is dissolved away by digestive enzymes to release a spherical onchosphere or hexacanth embryo, so called because it is armed with six curved hooks, with which it proceeds to bore its way through the wall of the stomach to reach the blood stream. When it arrives in muscle capillaries it bores its way out and settles between the muscle fibres. Here it loses its hooks and changes to a fluid-filled bladder, the bladder-worm or cysticercus. From one point on its wall, and projecting inwards, a scolex develops like the finger of a glove which is pushed from outside in towards the palm.

No further development will take place unless the joint of beef containing the bladder-worms is eaten by man without being sufficiently cooked to kill them. When one of these still living bladder-worms reaches its final host's intestine the scolex becomes everted from its previous position tucked into the bladder and comes to project from its surface. The bladder is now shed, having served its purpose, leaving the scolex to attach itself to the wall of the intestine and begin its life's work of budding off proglottides.

In the adult breeding tapeworm the eggs produced by one proglottis may be fertilised by the spermatozoa produced by the same proglottis, or by spermatozoa produced by another proglottis of the same tapeworm, which can easily occur because the worm will probably be doubled back on itself a number of times to fit its length into the intestine. There may even be cross fertilisation between proglottides of different individuals if there should be more than one tapeworm along the same length of intestine.

Although several phyla come under the general heading of roundworms, only two need concern us here. The more important of these is the phylum Nematoda, the nematode worms, an extremely important group of lower invertebrate animals containing many important parasites of man and his cultivated plants. The only other phylum in the group that we shall consider briefly is the phylum Nematomorpha. Only the expert with the microscope or the marine biologist is likely to come into contact with members of the other phyla.

In general organisation the roundworms represent little advance on the flatworms. The chief difference is that of shape, as the two names suggest. The roundworms, too, are completely devoid of external cilia, and in the majority the sexes are separate. As with the flatworms, there are both free-living and parasitic members.

Minute nematodes teem almost everywhere, especially in the soil and in the mud at the bottom of ponds and other stretches of still water. Although there is considerable variation in size, all the nematode worms are remarkably similar in general appearance, having cylindrical bodies that are pointed at both ends. If a small amount of the mud from the bottom of a pond is examined under a hand lens it will almost certainly reveal numbers of minute worms wriggling about in a side-to-side movement that is characteristic of all nematodes. The best way to obtain a corresponding sample from soil is to bury a small piece of meat in moist earth and leave it for a few days. It will attract dormant encysted larvae which burst into activity under the stimulus of the meat and begin to breed within a few days. If the meat is dug up after about a week, its surface examined under a hand lens will be seen to be seething with enormous numbers of tiny nematode worms and their larvae.

Because of the great economic importance of many of them much more is known about the parasitic than the free-living nematodes. The most widely distributed human nematode parasite is *Ascaris lumbricoides*. It is one of the largest of all the nematodes, the adults that live in the small intestine attaining a length in excess of one foot when fully grown. The males are somewhat smaller than the females, and mating takes place in the intestine. *Ascaris* females are extremely prolific, each female being capable of laying as many as two hundred thousand eggs a day, which pass out of the body with the host's faeces. Each egg before it is laid is surrounded by an extremely resistant shell, which can be dissolved away only by the digestive enzymes of the small intestine, and this means of course that no egg can complete its development unless it is first swallowed by its future host. At the same time these well-protected eggs are capable of remaining dormant for a number of years and still continuing with their development if they are then swallowed.

The first stages of development do occur, however, after the egg leaves the host's body but before it is picked up and swallowed by the next host, and while it is still retained within the shell. The first larva that results from the division of the fertilised egg can then moult to form a second larva. This is the infective larva, that is the larva that is adapted to settle in the human intestine, and for this reason there can be no further development within the shell until the egg is swallowed and the shell dissolved away.

One might expect that once this has taken place the released larva could settle down to grow to full size in the intestine and reproduce the next generation. But in fact a most unusual series of events follows. As soon as they are released these infective larvae burrow into the intestinal wall and reach the blood vessels within the wall. Within twenty-four hours they have travelled in the blood stream and have come to rest in the liver, where they stay for up to a week. They then resume their peculiar journey round the body, travelling on to the heart and then via the pulmonary arteries to the lungs. Within the lungs there is another pause while they undergo two more moults, after which they break through the walls of the lungs into the air sacs or alveoli. Having left the blood stream they now travel up the bronchi and trachea to find themselves at the back of the throat, from where they are swallowed and find themselves in the intestine for the second time. This time they do really remain there for the rest of their lives.

A careful search in almost any pond in the summer may well reveal some curious, thread-like, incredibly thin worms up to six inches in length. These are hairworms, belonging to the phylum Nematomorpha. They are quite common and widespread in fresh water. In earlier times they were thought to be horse hairs which had fallen into the water while the horse was drinking and had become animated! They are indeed similar in size and appearance to horse hairs, and they do appear quite suddenly without apparent reason in incredible numbers in ponds. Even today not all the facts of their life history are known.

Adult hairworms do not feed; their sole function is to reproduce. The females lay their eggs in long white strings which they twine around water plants. The larvae that hatch from them swim around until they encounter the aquatic larva of an insect, maybe a dragon-fly or a beetle. With the aid of a sharp proboscis with which they are provided they bore into the flesh of their host. From this primary host they somehow transfer to an adult land insect, such as a beetle or a grasshopper, though how they effect this transfer is at present unknown. Within the body of the land insect they develop into adults. They then leave the body of this last host and make their way back to water. One of the best known species is *Gordius robustus,* so named because many individuals often congregate in a tangled mass reminiscent of the mythological Gordian knot.

CHAPTER TEN

EARTHWORMS, MARINE WORMS AND LEECHES

In contrast to all the other creatures called worms, the members of the phylum Annelida have bodies that are constructed segmentally, the internal divisions between the segments being clearly indicated externally by circular constrictions of the body wall. There are two main classes within the phylum, the Chaetopoda and the Hirudinea. The first comprises two orders, the Oligochaeta, which consists of the earthworms and a number of freshwater worms, and the Polychaeta, an important group of marine worms, some of which are much used by fishermen as bait. The Hirudinea are the leeches, aquatic worms adapted to a life of external parasitism.

The earthworm leads an almost exclusively underground life, in burrows in the soil which it excavates for itself. On the lower surface of each segment there are four pairs of stiff, hair-like setae which enable it to grip. As it moves through the burrow the setae of the hind segments are thrust into the soil; while those of the front section are withdrawn while it is lengthened. The setae of the front end are now extended to anchor it to the burrow, while those of the hind end are withdrawn and the hind end itself is contracted, thus bringing it up towards the front end. By repeating this process the worm moves through its burrow in a forward direction, the reverse process being adopted to enable it to retreat.

The burrow itself is excavated by the simple process of eating the soil. Any organic material it contains is digested before it is passed out from the hind end of the body. In most species this digested soil is passed out below ground, but in some it is deposited on the surface as the familiar worm casts. Earthworms are essentially nocturnal creatures, coming to the surface only at night, when they anchor themselves to the mouth of their burrows with their tails and move around in search of fallen leaves. These are secured and dragged into the burrow to be used as food.

Earthworms are hermaphrodite, but they are not self fertilising. Their reproductive organs are complicated, as they are in the flatworms, and so also are the events that occur at and following copulation. The reproductive structures occupy segments 9 to 15 and they are all paired, one member of each pair being on each side of the mid-line.

There are two pairs of testes, one pair situated on the front wall of segment 10 and the other pair in a similar position in segment 11. In the mature worm the testes are covered by lobed chambers, the seminal vesicles, which extend from the 9th to the 12th segments. The spermatozoa are shed into these seminal vesicles where they complete their development. Two pairs of funnels open forwards from the hind walls of segments 10 and 11 into the cavities of the seminal vesicles, and the ducts from the two funnels of each side open into a single vas deferens which leads backwards to open on the under-surface of the 15th segment.

The female reproductive organs are somewhat less complicated. There is a pair of ovaries on the front septum of segment 13 and a pair of ovisacs on the hind wall in which ripe eggs can be stored until the time comes for them to be released. Beneath this pair of ovisacs is a pair of funnels leading into the two oviducts, which open beneath the 14th segment. Quite separate from these male and female reproductive organs are two pairs of blind sacs, the spermathecae, opening on the under-sides of segments 9 and 10. These receive the spermatozoa of the partner during copulation and retain them until they are required for fertilising the eggs.

Mating takes place in the spring, when the worms come together in pairs, the head of one lying towards the tail of the

Earthworms mating

other. There is a copious secretion of mucous in the region of the reproductive organs so that the two bodies in this region are surrounded by a mucous tube, under cover of which the spermatozoa of each individual are passed out through the vasa deferentia opening on segment 15 and find their way into the spermathecae of the partner opening on segments 9 and 10. This exchange of spermatozoa having been completed, the two worms separate, the actual fertilisation of the eggs taking place after the completion of copulation.

In the next stage of reproduction the saddle or clitellum plays a leading part. This glandular area of the earthworm's body occupies a variable number of segments between segments 30 and 40, the actual number depending upon the species. Some time after copulation the clitellum secretes a cocoon, into which it also secretes some albuminous fluid.

This will serve to nourish the embryos that the cocoon will eventually contain. The cocoon is moved forward by muscular activity of the body wall. As it passes segment 14 the ripe eggs are passed out into it from the oviducts, and then as it continues its forward movement it receives the spermatozoa previously obtained from the mating partner from the spermathecae opening on segments 9 and 10. Having now received eggs, spermatozoa and albuminous fluid the cocoon continues its forward movement until it clears the head of the worm, when it becomes sealed off both in front and behind to form an egg capsule, in which the fertilised eggs develop directly to young worms, which eventually escape from the capsule into the surrounding soil to lead an independent life.

In addition to the earthworms, the order Oligochaeta also contains a considerable number of worms that live in fresh water. One of the most interesting of these is *Stylaria,* a tiny transparent worm generally less than half an inch in length and usually to be found crawling on water weed. It has extremely long bristles and a long slender proboscis projecting from the front end. Because of its lack of colouring most of its organs can be seen under a powerful hand lens. *Stylaria* is a member of the family Naididae, and in this family sexual reproduction is not very common. The more usual method of reproduction is asexual reproduction by budding, new individuals being produced by the constriction of the hind end of the body. These asexually produced offspring may remain attached to the parent for some considerable time while others are in their turn budded off behind it, so that what at first appears to be a single worm may in fact be a long chain of new individuals still attached to the original parent.

In contrast to the members of the order Oligochaeta, which are either terrestrial or fresh water, the Polychaeta are all marine or seashore worms. The chief distinction between them and the oligochaetes is that, whereas the latter bear their setae on the surface of the segments, in the polychaetes they are borne on special paired fleshy outgrowths on each segment. These parapodia, as they are called, have the appearance of flat paddles. Polychaetes are typically active free-living worms with powerful jaws which they use to capture their prey. They are the ragworms so much prized as bait by the sea angler, who searches for them in their hiding places under stones and in rock crevices.

In the polychaetes, in contrast to the oligochaetes, the sexes are usually separate, but their method of reproduction is unusual. The reproductive products in both sexes are formed in the segments towards the hind end of the body, and as they ripen so the segments in which they are developing undergo considerable changes. The parapodia become enlarged and flattened. At the same time the colour of the worm darkens and the eyes grow larger. When these changes have been completed the worms leave their hiding places and swim about in the sea, using their enlarged parapodia as paddles. Soon the body wall of their hinder segments splits and the reproductive products are shed into the sea, where the eggs become fertilised. Tiny larvae hatch out from these fertilised eggs, and after a period of planktonic life they settle on the sea bed and change to young worms. So different are the worms at the breeding season that for a long time the breeding individuals were thought to be different species, and accordingly were separately named.

The larva that develops from the eggs of these marine polychaetes is known as a trochophore. It has bands of cilia which enable it to swim during its planktonic existence. This trochophore larva is extremely important from the evolutionary point of view, because it is very similar to the larvae produced by members of several other phyla, and thus gives a clue to the relationship between these phyla.

One of the most remarkable of all these polychaetes is the Pacific palolo worm, *Leodice viridis*. The story of the palolo worm is one of the best examples of the dependence of animal life upon the changing seasons. In the seas around many of the tropical Pacific islands this worm appears in enormous swarms near the surface on four days in the year. At other times not a single specimen will be found. To the native the islands the worms are a delicacy and they catch gre numbers during these four days, ladling them out of the sea by the basketful. Long ago it was realised that the palolo worms always appeared on exactly the same four days. Their first appearance is on the morning of the day before the last quarter of the October moon, and they persist throughout the next day. By the third morning they have all disappeared. A month later, again on the day before the last quarter of the moon, they make another appearance for two days. These November swarms are larger than the October ones. After this second appearance nothing is seen of the palolo worms until the following October.

So much must have been known from very early times, for the Fiji natives have called these two periods in their calendar *Mbalolo lailai* (palolo little) and *Mbalolo levu* (palolo large). Modern investigation has revealed the constancy of swarming as only a part of a remarkable life history. The little worm lives through the year hidden among corals or in rock crevices. As October approaches the hind end undergoes drastic changes: numerous pairs of paddle-shaped structures appear along each side while the reproductive products are ripening within. Then, at dawn on the morning before the moon's last quarter, either in October or November, the worm backs out of its hiding place until the hind part is free, whereupon it is shaken off and starts paddling for the surface, in the company of myriads of its fellows. Meanwhile the front end retires to its crevice for another year. Once at the surface the worms paddle around for a time before disintegrating to release their reproductive products. Then their spent bodies sink exhausted to the sea bed, where they provide food for other animals.

A species quite closely related to the palolo worm, *Leodice fucata*, is found on many West Indian reefs, and it too shows a remarkable correlation between its time of reproduction and the state of the moon. Spawning occurs only during the third quarter of the June–July moon.

In addition to the two main annelid classes there are a few

much smaller classes whose members are little known. One member of one of these classes must however be mentioned because of the curious relation between the sexes. The worm is *Bonellia viridis*, a member of the class Echiuroidea. The female has a plump ovoid body about two inches in length with a long thin proboscis forked at its distal end growing out from its front and just above the mouth and attaining a length of some six inches. It is of an attractive green colour and lives in rock crevices along the warmer coasts of Europe.

By contrast the male is a tiny and rather degenerate creature about one sixteenth of an inch in length and covered with cilia. It has no digestive system and is incapable of leading an independent existence, spending the whole of its life as a parasite on the body of the female. At first it attaches itself to the proboscis, but as it grows it migrates to the cavity of the mouth. Finally, as it becomes sexually mature, it migrates for a second time and enters the opening of one of the female's nephridia or excretory pores.

Both sexes shed their reproductive products into the sea, where the eggs become fertilised. The sex of the resultant larvae is apparently not fixed at fertilisation as it is with most animals. If a larva develops independently it will give rise to a female, but if it should develop in contact with the body of a female its development is greatly inhibited, and it will produce a dwarf parasitic male.

The leeches, comprising the class Hirudinea, are an interesting group of annelids which have become adapted in structure and modified in behaviour to a life of parasitism. They are all, however, free-living creatures, attaching themselves to their particular host species only when a meal is required, sucking in sufficient blood or body fluid before releasing their hold.

At first sight a leech does not bear much resemblance to an earthworm. It is relatively shorter, its body is flattened, and there are no setae. There are however surface rings, which look very similar to those of an earthworm but are not external indications of internal segmentation as they are in the earthworm. The leech body is segmented internally, but several of these external rings correspond to each of the true segments. One feature, however, is shared by both groups. An individual leech, like an earthworm, can extend or contract its body to a very considerable extent.

The most characteristic external features of any leech are its two suckers, one at either end of the body. The one at the head end is very small and often not easy to see. It surrounds the mouth, and its purpose is to fix the mouth firmly to the host's body while it feeds. The much larger hind sucker serves to anchor the body to a suitable support at all times. Different kinds of leeches have two different methods of piercing the skin of their victims. In one group a circular proboscis penetrates the host's body, while in the other the mouth is armed with three jaws which inflict a Y-shaped wound.

Leeches, like earthworms, are hermaphrodite, but so far as is known self fertilisation does not occur. The method of transfer of the spermatozoa at mating is, however, different from any we have met so far. Before being passed out from the vas deferens they are grouped together in lance-shaped capsules which are thrust into the skin of the mating partner. From these spermatophores, as the capsules are called, the spermatozoa are released into the body tissues, and then make their way through these tissues until they reach the ovaries, where of course they fertilise the eggs.

Most leeches lay their eggs in cocoons as earthworms do, and in these species a clitellum develops towards the front end of the leech at the breeding season. The fertilised eggs are deposited in the cocoon as it passes the external opening of the oviduct. In most species the cocoons are attached either to stones or submerged plants. The horse leech, *Haemopis sanguisuga*, however, leaves the water after mating and deposits its egg cocoons in damp earth nearby, and the young leeches that hatch out make their way back to the water. Horse leeches do sometimes leave the water for short periods at other times. The land leeches of tropical and subtropical countries are related to the horse leech.

Members of the family Glossosiphonidae are different in their reproductive methods from all other leeches. Their eggs, instead of being deposited in cocoons, are laid singly. In some species they are attached to submerged plants or other suitable supports, where the parents keep watch over them until they hatch, and in others they are attached to the under-side of the parent body. In both cases the young leeches that emerge from the eggs attach themselves to the under-side of the parent body, where they remain until they have grown sufficiently to be able to leave the protection of the parent to fend for themselves. More than two hundred young leeches have been counted attached to a single parent.

CHAPTER ELEVEN

MOLLUSCS

The phylum Mollusca consists of a great variety of soft-bodied animals typically with a protective shell covering much or all of the body. There are five classes, of which three are common and widespread, and which differ very considerably in fundamental structure. The class Gasteropoda comprises the snails and slugs, which are common in the sea, in fresh water and on land, the majority being instantly recognisable by their prominent coiled shells. The class Lamellibranchiata includes all the bivalve molluscs—the oysters, mussels, clams, cockles, scallops and a host of others—which have a more or less flattened shell consisting of two distinct halves or valves hinged together to form a box capable of enclosing all or at least part of the body. It comes as a surprise to most people to learn that the octopus with its eight arms and the squids and cuttlefish with their ten arms are also molluscs comprising the class Cephalopoda.

The body of a typical gasteropod consists of three main parts, a head bearing eyes and sensory tentacles and behind it a thick flat muscular foot on which it can walk. Above the foot is the visceral hump, containing the gut and all the other internal organs and covered with a curtain of tissue known as the mantle. At the front end of the snail the mantle hangs free of the hump to form a mantle cavity between the two. The main function of the mantle is to secrete the calcareous shell which covers the hump and into which the head and foot can be withdrawn if danger threatens. The visceral hump is twisted or coiled, and this explains why the shell laid over it is also coiled. Because of this twisting, too, the gut does not open at the hind end, as it does in most animals, but into the mantle cavity above the head. It is believed that the ancestral gasteropod had an untwisted body in which the mantle cavity was at the hind end, with the anus opening into it, the twisting or torsion, as it is called, having been developed at some time during the course of subsequent gasteropod evolution. The larval development of modern gasteropods provides important evidence to support this belief, as we shall see.

Most of the more primitive marine gasteropods, like for example the common limpet, *Patella vulgata,* adopt the broadcast method of breeding, shedding their reproductive products into the sea at the breeding season. In the case of the limpet this occurs during the winter, so that larval development is completed by the spring and the young limpets are then ready to settle out of the plankton. In general only those that settle on a clear rock surface are able to survive, and most of the remainder, which means the vast majority, are destined to perish.

Recent work has shown that in many marine invertebrate species it is possible for larvae that initially settle on unsuitable ground to migrate for short distances during a kind of swimming–crawling stage. In this way they are given a second chance of selecting a suitable substratum where they will be able to continue their development to adults. This intermediate trial period will certainly not last for more than a week, and only those larvae that initially settled relatively near to suitable ground will stand any chance of a reprieve.

As soon as the limpet egg has been fertilised it develops into a ciliated trochophore larva remarkably similar to the trochophore larva produced by the marine polychaetes. After a time the trochophore develops into a second type of free-swimming larva known as a veliger, in which a special additional ciliated zone known as the velum develops. This may take the form of a circular plate or of a number of ciliated lobes.

It is during the veliger phase that the rudiments of the adult organs are laid down, including the shell and the foot.

When the gut is first formed the mouth is at the future front end and the anus at the hind end. At some stage, however, a remarkable upheaval takes place. The whole of the visceral hump proceeds to twist through a full one hundred and eighty degrees, thus bringing the anus and the mantle cavity to the front where it comes to lie immediately above the mouth and the developing head. This whole process of torsion can take place remarkably quickly, having been observed under laboratory conditions to occupy no more than a couple of minutes.

In most of the more advanced marine gasteropods the eggs are fertilised internally after mating and are covered with a protective capsule before they are laid. The trochophore stage is usually passed within the capsule, and the larva eventually emerges as a veliger. In some species even this stage occurs within the capsule, the animals emerging as young adults.

The four species of periwinkles that can be found in abundance on most rocky shores show interesting variations in their methods of reproduction. The common periwinkle, *Littorina littorea,* lives on the lower parts of the shore, and after mating the eggs are laid in helmet-shaped capsules containing up to four eggs. In these capsules the trochophore stage is completed and the larvae that emerge have reached the veliger stage. These veliger larvae live for a time in the plankton before settling as young periwinkles to assume adult life if they are fortunate enough to settle on a suitable substratum.

In the middle regions of the shore two species are found, the rough periwinkle, *Littorina rudis,* and the flat or blunt periwinkle, *Littorina littoralis.* The wastefulness of the broadcast method of breeding among marine animals has already been made abundantly clear. The flat periwinkle has eliminated the waste inherent in a planktonic larval existence, and has thus drastically reduced the number of offspring it needs to produce. It lays its eggs enclosed in cases on the seaweed fronds among which it lives, and the eggs undergo complete development within these cases. Only when the larval stages have been completed do the complete young periwinkles emerge to grow to adults.

The rough periwinkle has gone one stage further in protecting its larval stages from possible predators. The fertilised eggs are retained within the body of the female snail while they undergo their larval developmental stages, and only emerge as young adults already possessed of a tiny shell. Because of their protective devices the flat and rough periwinkles produce many fewer eggs than the common periwinkle.

The small periwinkle, *Littorina neritoides,* has become almost a land animal. It lives on rocks right at the top of the shore, where it is splashed with sea water only for a few days each fortnight at the spring tides. Its breeding activity is modified to fit in with this intermittent contact with the sea. After being fertilised internally the eggs are enclosed in capsules and shed into the water. Reproductive activity, however, occurs only once a fortnight over a period of several months during the winter, this activity coinciding with the spring tides. At any other time it would be impossible for the egg cases to be liberated into the water.

The tiny larvae that hatch out from the eggs remain floating in the plankton for a considerable time. Those that survive the hazards of a planktonic existence will eventually settle down on the sea bed, most of them where conditions are unsuitable for their survival. Only those that settle on weed-free rocks on the middle to the upper parts of the shore will be able to develop. After a time they will crawl up the shore to make their permanent home in the splash zone.

The periwinkles are vegetarian, feeding on seaweed sporelings which they scrape off the rocks. Another common seashore snail is the dog-whelk, *Nucella lapillus.* This is a carnivore, able to feed on acorn barnacles and mussels. It is rather larger and has a thicker shell than the common periwinkle, and exists in a considerable variety of colouring. There are white, yellow, pink, pinkish-mauve and black individuals, as well as those striped with any two or more of these colours.

Breeding takes place in the early part of the year, mating taking place before the female lays her eggs. These she produces in batches each containing several hundred eggs, enclosing each batch in a small yellow egg capsule about the size and shape of a grain of wheat. Laying the eggs and constructing the capsule takes about an hour, but each female may produce two dozen or more batches at one spawning, and in a series of successive spawnings she can produce up to three hundred capsules. The capsules are attached close together to seashore rocks, usually in an upright position and sheltered in a shallow crevice or beneath an overhanging ledge.

Only about a dozen or so of the eggs in each capsule are fertilised. The remainder represent a store of food for the few that eventually hatch out after a prolonged development period of three to four months. Both larval stages are completed within the capsule, so that the tiny individuals that emerge are fully shelled miniature replicas of the parents that produced them. These young snails are carried by the receding tides down to the lower parts of the shore, where they live until they have grown to about a third of an inch in length. Only then do they crawl up to the middle parts of the shore where they will live out their adult lives.

Gasteropods generally are mobile animals, even though, like the common limpet, they may return to a particular spot after foraging trips. The slipper limpet, *Crepidula fornicata*, is unusual. It is not only sedentary, but has a habit of living in piles, one animal sitting on the back of another until a more or less permanent chain of up to nine individuals is formed, the lowest animals being the oldest. Its method of feeding, too, is quite unlike that of any other gasteropod. Instead of eating seaweeds or other animals, it is provided with much enlarged gills which trap the minute plankton organisms suspended in the respiratory water current, a method typical of the bivalves, as we shall see.

The lowest animals in the chain are always females, while the top two or three, the youngest in the chain, are males. Those in between are intermediate, in process of changing from male to female. The reproductive products are shed into the sea, where a complete larval development takes place before the young adults eventually settle.

Gasteropods were evolved in the sea. Eventually some types, which had become adapted to living on the shore, where for certain periods each day they were exposed to the air, gave rise to types that were completely independent of the sea. And so the first land snails appeared. Subsequently some of these land snails evolved further and gave us some of our freshwater snails by becoming readapted to living in water, although now it was fresh and not salt water. Parallel with this secondary return to water, other groups of marine snails migrated from the sea through river estuaries until they, too, became adapted to life in fresh water. Thus our modern freshwater snails have two distinct origins which are reflected in certain aspects of their structure. Those that have been evolved from land ancestors belong to the sub-class Pulmonata, while those whose ancestors were marine belong to the sub-class Prosobranchia.

The best known of the freshwater pulmonates are the pond snails, all species of the genus *Limnea,* the largest being the great pond snail, *Limnea stagnalis,* which does sometimes occur in canals and slow flowing rivers as well as in ponds. It lays its eggs after they have been fertilised in sausage-shaped masses of jelly attached to aquatic plants. There is a curious connection between the size of the snails and the size of the pond in which they are living. Only in a very large pond will they grow to full size, the maximum size achieved in smaller ponds varying according to the size of the pond.

This correlation between the size of the animal and the amount of water it has to live in has been demonstrated experimentally. Young snails hatching from a single batch of eggs were separated into several groups, which were then reared in widely different volumes of water. Each batch was provided with ample supplies of food. The results showed that not only the maximum size, but the rate of growth as well, was dependent upon the total volume of available water.

The dwarf pond snail, *Limnea truncatula,* is a much smaller species, seldom attaining half an inch in length, but otherwise similar in appearance to its large relative. Despite its name it is more a marsh than a pond snail, being found typically on swampy pasture. Its eggs are laid out of water on mud or on dead leaves, in little gelatinous packets each containing about a dozen. Laying starts in March, and since the young snails emerge in about three weeks, and can begin breeding themselves in another eight to ten weeks, at least two generations can be completed in a single season.

The dwarf pond snail's chief claim to fame, or at least notoriety, is that it is the intermediate host for the larval stages of the sheep liver fluke. Until the life history of the fluke was known the annual losses of sheep through liver fluke attacks were immense. The discovery of the part played by the snail enabled farmers to devise effective remedies, the most important being to drain damp pastures to make them too dry for the snails, and to keep sheep off pastures that cannot be drained.

Pulmonate freshwater snails are found in every type of fresh water, as one might expect, and equally predictably the prosobranchiates are found only in streams and rivers. It is difficult to see how they could have crossed land barriers to colonise ponds. In their methods of reproduction most of them resemble the pulmonates, and attach their egg capsules to some suitable substratum from which they cannot easily be swept away. The freshwater winkles or river snails, however, are viviparous; the fertilised eggs are retained within the oviduct while they develop, and the young snails are not released into the water until they are fully formed.

Mating in land snails and slugs is quite an elaborate affair. All of them are hermaphrodite, but cross fertilisation is the rule. At the mating season in the spring they come together in pairs and indulge in mutual copulation. Before this, however, there is a preparatory courtship accompanied by mutual stimulation. Within the female duct or vagina there is a special sac called the dart sac, in which a calcareous spicule known as the love dart is formed.

When a pair of snails or slugs come together for mating each aims this dart through the opening of the vagina and into its partner's flesh with sufficient force to carry it right through

Snails involved in mutual stimulation

the body wall as far as the internal organs. This rather drastic greeting stimulates both participants. They now approach closely, each inserting its own penis into its partner's vagina, into which the spermatozoa are passed enclosed in a chitinous envelope called a spermatophore. This envelope is eventually dissolved away to release the spermatozoa, making them available for fertilising the eggs when they are laid. Egg laying occurs in early summer, and they are buried in the ground, where they will be protected both from enemies and from desiccation.

The lamellibranchs differ very considerably in their structure and habits from the gasteropods. Pehaps the best known of them all is the common mussel, *Mytilus edulis*. Its structure is typical of the lamellibranchs. Within the two hinged shell valves the mussel's body is covered with a mantle consisting of a right and left half, each responsible for forming one of the two valves. The pointed end of the shell is the front end. The two valves are closed by the contraction of a pair of adductor muscles, one in front and one behind, running across from one valve to the other. Towards the top the mantle is fixed to the body, but below the frilled edges of the two halves remain free. Behind they are partially joined together, leaving two openings one above the other. These are the inhalant and exhalant siphons referred to below.

If one of the valves is turned right back after the adductor muscles have been cut away from it, the structure of the body can be seen. Inside the mantle flap there are two other large flaps of tissue on each side, running practically the whole length of the body. These are the gills. Between them towards the front of the animal a foot projects downwards. When the valves are open it can protrude beyond the shell, but except in the very young mussels it is not normally used for movement. Its function is to produce those tough threads by which the mussel attaches itself firmly to the rock. These are the byssus threads, and they are produced as a sticky liquid by the byssus gland in the foot. When required, the liquid runs down a groove on the foot, which is used to place it on the rock. As the foot is withdrawn, a thread is formed which hardens almost at once. In a short time a complete series of byssus threads can be produced, sufficient to keep the mussel firmly attached to the rock even in the heaviest seas. You can test the strength of the byssus by trying to pull a living mussel away from the rock to which it is attached. The byssus gland is probably a modification of the gland in the snail's foot which produces the slime on which it walks.

In contrast to the well developed head of the gasteropods, the bivalve is practically headless. It has no tentacles or sense organs in the head region, only a mouth surrounded by flaps called palps which are used to sort out its food. The only sense organs it possesses are small sensory tentacles on the hind edge of the mantle.

As you see them covering the rocks in their thousands when the tide is out they are inactive, with their valves tightly closed. It is only when the tide comes in and they are covered with water that they relax their muscles and allow their valves to open. A current of water passes in through the lower inhalant siphon, travels through the numerous minute channels perforating the gills and leaves the body by the upper exhalant siphon. In many bivalves the edges of the siphons are developed into tubes which may project out beyond the shell.

This water current is not caused by muscular action, but by the rhythmical beating of countless minute hair-like cilia covering the surface of the gills. It is of course a respiratory current, providing a continual supply of fresh water for respiration, but it also provides the animal with its food. The gill perforations are so small that even the microscopic unicellular plants called diatoms are too large to pass through them, and become trapped on the surface. They are taken up by a sticky mucous produced by some of the surface cells, and conveyed forward to the mouth by another set of cilia. The palps now sort out the food and allow only diatoms and small particles to enter the mouth. Larger rejected particles leave the body with the exhalant current, as do the undigested remains of the food after it has passed through the gut.

Suspension feeding by filtering out the plankton suspended in a current of sea water is adopted by most lamellibranchs as well as by members of other phyla. The advantages of this method of feeding, with each tide bringing fresh supplies of food, must be considerable, judged by the great concentration of mussels and oysters that can thrive on a good bed. The amount of water filtered at each tide by a heavy population of mussels or oysters must be very great, many pints passing into and out of each individual. Figures for the total rate of growth on mussel beds certainly underline the potential efficiency of suspension feeding. It has been calculated that on a flourishing mussel bed the total increase in weight in a year may amount to forty thousand pounds per acre, representing ten thousand pounds of actual mussel meat. An acre of good pasture, on the other hand, is calculated to be able to produce between one and two hundred pounds of beef every year.

Although the majority of lamellibranchs live buried in sand or mud, they adopt exactly the same methods of respiration and feeding. When they are covered by the incoming tide they lie just below the surface with the two siphons projecting just above the surface of the sea bed, and they can also exist in enormous concentrations. On a really good cockle bed, for example, it has been calculated that there may be as many as one and a half million cockles buried beneath the surface of an acre of muddy sand.

Reproduction is similar for the vast majority of marine and seashore lamellibranchs. At the breeding season in the spring the reproductive products are shed from the reproductive organs into the mantle cavity, finally reaching the exterior with the exhalant current. The trigger that sets off spawning activity is the temperature of the water. When this reaches a certain minimum figure, which varies according to the species, a few individuals begin to spawn. The presence of their reproductive products in the water in some way stimulates all the other individuals in the same area to spawn, so that within a short time the whole population is spawning. Such a mechanism of mutual stimulation is an obvious advantage where random fertilisation without mating has to be relied upon, because it produces maximum concentration of reproductive products in the water, and therefore ensures a high proportion of fertilised eggs.

The duration of larval life varies considerably in the molluscs. In some species the larvae spend only a short time in the plankton, seldom more than seven days, before settling. Most lamellibranchs and some gasteropods, however, have an extended larval life, the veligers remaining in the plankton for two or three months, during which time they will be distributed very widely. During this time they are able to feed and grow on minute food particles trapped by their bands of cilia and conveyed to their larval mouths. *Mytilus edulis* can produce up to twelve million eggs at a spawning, each capable of developing into a veliger larva.

As a result of their great reproductive and distributive

capacity virtually any suitable surface that becomes available will be colonised. A striking example of their colonising efficiency occurred in Holland towards the end of the last war. When the German troops blew up the dykes surrounding the island of Walcheren in 1944 the land behind them was of course flooded with sea water. In the following year the dykes were repaired and the water pumped out. During the year that they had been covered with water, mussel larvae had settled on the walls of houses, on fences and even on trees, and these were found to be covered with growing mussels.

Freshwater mussels cannot adopt the same method of reproduction as the marine mussel because prolonged larval stages living free in the water of a river would be swept out to sea. During the summer breeding season the females produce thousands of eggs which are retained in special brood pouches situated among the gills. The males, however, shed their spermatozoa into the water, and these are eventually drawn into the females in their respiratory currents and so fertilise the eggs. These fertilised eggs remain in the brood pouch right through the winter, developing slowly. Eventually, the following spring, they hatch out as special larvae known as glochidia. Each glochidium has a pair of tiny triangular-shaped shell valves, each bearing a single sharp tooth at its free apex. As the water becomes warmer these larvae are released and float about in the water for a short time before sinking. A single sticky byssus thread projects from the foot between the open valves, and serves to attach the larva to any plant or other object with which it comes into contact.

Further development is impossible as free larvae. In order to survive a glochidium must become attached to a fish and continue its development as a parasite. If a fish passes close to it it may become attached to it by its byssus thread and be carried away. By means of its shell teeth the glochidium buries itself beneath the skin of the fish, usually on the fins, tail or gills. The tissues of the fish, irritated by the unwanted guest, react by forming a cyst around it, and in this protective coat it proceeds to develop.

Of course very few of the glochidia are fortunate enough to be picked up in this way. The unfortunate majority soon die, but the successful ones continue their development for about three months, after which time they are ready to leave their hosts and drop to the pond or river bed to continue their life as adult mussels. All kinds of freshwater fish serve as hosts to the mussel larvae, but the three-spined stickleback seems particularly favoured. Sometimes a single specimen may show several dozen swellings on its fins and tail, and each one of these indicates the presence of a developing glochidium.

In most other freshwater bivalves the fertilised eggs are retained in the gill pouches of the females until they are fully formed young adults, able to settle on the pond or river bed as soon as they are released.

Only the zebra mussel, *Dreissena polymorpha,* produces veliger larvae similar to those produced by the marine mussels, and these settle on the river bed after a week or two of larval life. They are able to exist only in slow-flowing rivers—in swifter streams the larvae would be swept out to sea before they had had time to sink and anchor themselves to the river bed.

The vast majority of living Cephalopoda belong to one of two orders, the Decapoda and the Octopoda. The decapods are the squids and cuttlefishes, which have ten arms, eight short ones and two very long ones that only have suckers at their tips. They are mostly active swimmers, and their bodies are streamlined. The octopods, which are the octopuses, are less active creatures and tend to spend most of their time on the sea bed, though they can swim if necessary. They have eight arms, all of similar length.

Both groups, however, adopt essentially the same method of reproduction. The sexes are separate, and mating is preceded by a period of courtship which in some species can be quite elaborate. This culminates in the male depositing one or more spermatophores in the mantle cavity of the female, the spermatozoa that they contain being used to fertilise the eggs which are laid sometime later. At the breeding season both in octopods and decapods one arm becomes modified for spermatophore transfer, this arm being known as the hectocotylus. The suckers towards its tip are used to extract the spermatophores from its own reproductive organs and deposit them in the female mantle cavity.

Cephalopod spermatophores are quite complicated structures produced by a very specialised male duct, of which each male possesses only one. They have been most intensively studied in the small squid *Loligo*. Each spermatophore is shaped like a thin torpedo about two-thirds of an inch in length. About two-thirds of its length is occupied by a mass of spermatozoa embedded in mucous and surrounded by an inner tunic. At its narrower end next to the remainder of the spermatophore the inner tunic is capped by a small cement body, to which is attached in turn a spiral filament which continues to the end of the spermatophore. The whole structure is enveloped in a second outer tunic, the end of which consists of a terminal cap from which a thin thread is drawn out.

A male *Loligo* can produce up to a dozen spermatophores in a day during the breeding season, and in a special pouch called Needham's Sac it can store as many as four hundred during the breeding season ready for use. At mating the hectocotylus arm pulls a bunch of spermatophores out of the sac and transfers them to the female's mantle cavity. As they are detached the terminal caps are loosened. In the female mantle cavity the cement body is forced out of the outer tunic and becomes attached to the mantle wall, which ensures that the sperm mass is securely fixed within the female.

When they are eventually laid and fertilised the eggs of cuttlefish and squids are usually fixed in strings or clusters to stones or vegetation, often on special spawning grounds where large numbers of individuals deposit their eggs. Once they are laid the females take no further interest in them. Cuttlefish eggs are enclosed in tough black capsules, giving them the appearance of small black grapes and protecting them from potential predators. Squid eggs are enclosed in jelly which is distasteful to most other animals.

In contrast the female octopus looks after and protects her eggs until they hatch. She usually deposits them in some rock crevice and cradles them in her arms, shooting jets of water over them to keep them aerated. Squid, cuttlefish and octopus young live for some time in the plankton after hatching before becoming young adults.

The most remarkable of all cephalopod mating is that of the argonauts or paper nautiluses. These are octopods that have retained the ancestral molluscan shell. The female common nautilus measures up to six inches across its body and twelve inches across its shell. For a long time only females were known, but eventually the male was discovered, and was found to be minute, only about one-twentieth the size of the female. But at the breeding season he did develop a relatively enormous hectocotylus up to five inches in length. This hectocotylus detaches itself from the male and can swim free in the sea until it meets up with a female, to which it attaches itself and transfers the spermatophores that it carries. These detached arms were known to the great French anatomist Cuvier, who regarded them as parasites of the argonaut, and gave them the name *Hectocotylus octopolis*, from which we get our present name hectocotylus for the mating arm of the male cephalopod.

In these argonauts the fertilised eggs are also cared for in a remarkable way. Two arms secrete a fragile coiled calcareous shell in which the eggs are protected until they hatch. These shells are of course not homologous with the true mollusc shells.

CHAPTER TWELVE

CRUSTACEANS

The twenty or so phyla or major groups into which the animal kingdom is divided are very unequal both in the number of species they contain and in the total number of individuals belonging to them. Some of them contain relatively few species confined in their distribution, whereas others are represented by an abundance of types and individuals in almost every part of the world. If such abundance is to be the criterion on which the success of a phylum is to be judged, then there is no doubt that the most successful phylum in the animal kingdom is the Arthropoda. More than three-quarters of all the animal species at present known to science belong to it, and the number of individuals that can be allocated to it vastly exceeds the remainder of the animal kingdom. The phylum is divided into five subphyla, and a summary of the main groups of animals included in these will give an impression of the vast scope of the phylum as a whole.

The subphylum Crustacea contains the shrimps, prawns, crabs, lobsters, crayfish, water fleas, barnacles, woodlice and many related animals. The majority of the members of the other four subphyla are terrestrial animals, whereas the crustaceans are predominantly aquatic, the majority of them being marine, with a minority of freshwater and terrestrial forms.

The next two subphyla, the Chilopoda (centipedes) and the Diplopoda (millipedes) used to be classed together in a single subphylum, the Myriapoda, but in recent times it has been realised that their superficial similarities, in particular their possession of large numbers of legs, were less fundamentally important than their differences, and that they were not as closely related as a cursory glance might indicate. Neither is a large group, though they are fairly widely distributed.

The subphylum Arachnida includes all the spiders, harvestmen, mites, ticks and scorpions.

Finally there is the subphylum Insecta, by far the largest group in the whole of the animal kingdom, containing well over three quarters of a million species.

The most important feature of the arthropods is that the whole surface of the body is covered with a cuticle which constitutes an external skeleton, the muscles that serve to move the body being internal to it and attached to it. The cuticle consists fundamentally of a flexible horny material called chitin, which is often strengthened and made rigid by being impregnated with other substances, usually calcium salts. On the outside it is covered with a thin layer of waxy material which is impermeable to water and gases. Where moveable joints are needed both on the body and on the legs the cuticle remains unstrengthened and so retains its primary flexibility.

The arthropod body is segmented, but at the front end in many types the segmentation is masked by cuticle thickening. Fundamentally each segment is provided with a pair of appendages, which are variously modified to form antennae or antellules with sensory function, jaws and other mouth parts for dealing with food, walking legs for locomotion either on land or on the sea or river bed, swimming appendages for travel through water, copulatory organs and organs to which fertilised eggs can be attached and so be protected while they develop, or, in some aquatic Crustacea, gills whose function is to extract the required oxygen from the water in which they live.

In Crustacea, Diplopoda and Insecta the body consists of three distinct sections, the head, thorax and abdomen. The number of segments contained in each may vary even within the same subphylum. The members of the Chilopoda have a distinct head but the rest of the body cannot be subdivided into a thorax and abdomen. In the Arachnida the head and thorax are fused together to form a composite cephalothorax, followed by a distinct abdomen.

The great disadvantage of a rigid external skeleton is that once it has been formed the animal within it cannot grow in size. Consequently it must be shed periodically and replaced by a new and larger cuticle.

In most Crustacea the sexes are separate. The typical larva is known as the nauplius, which has an egg-shaped body with three pairs of appendages, antennules, antennae and mandibles. In many types the nauplius is succeeded by one or more, and sometimes many, subsequent larvae before the adult form is finally achieved. The fertilised eggs, too, are often retained by the female in special brood pouches or attached to her appendages, and are only released to fend for themselves at an advanced larval stage.

The crustaceans are divided into six classes which are very unequal in the number of species and individuals they contain. The majority of the class Branchiopoda live in fresh water, only *Artemia*, the brine shrimp, and a few other marine forms being found in the sea. The most important members of the class are the fairy shrimp *Chirocephalus*, *Apus*, and *Daphnia* and its relatives, the water fleas.

The class Ostracoda is a small group of tiny crustaceans whose external skeleton has become modified to form a bivalve shell superficially similar to that of a bivalve mollusc and completely enclosing the body. *Cypris* is a common freshwater form and *Cypridina* an important marine representative.

The class Copepoda is an extremely important group of smaller crustaceans, its representatives forming the majority of the crustaceans found in the plankton both in the sea and in fresh water. *Calanus* is one of the most important marine representatives, and *Cyclops* one of the most widespread freshwater forms. The class also contains a number of common fish parasites.

The class Branchiura is a small group of temporary fish

parasites known as carp lice. They are able to swim quite strongly, but attach themselves to the gills of fishes, where they feed on blood drawn from the gill filaments. *Argulus* is one of the best known of these fish parasites.

The members of the class Cirripedia are extremely modified crustaceans, the adults of most species bearing no apparent resemblance to crustaceans at all. Their true affinities are revealed only through their development from normal crustacean larvae. They are all sedentary creatures as adults, and the majority of them are hermaphrodite. The ship barnacle, *Lepas*, the acorn barnacles, *Balanus*, etc, and the parasitic sac barnacle, *Sacculina*, which parasitises crabs, are the important members of the class.

The final class, the Malacostraca, includes all the higher crustaceans, the true shrimps and many other shrimp-like creatures, the prawns, lobsters, crabs, crayfish, woodlice and their many relatives.

The Branchipoda are the most primitive of all the living crustaceans, and unlike the other classes the majority of the members are freshwater creatures. Particularly interesting in its habits is the tiny fairy shrimp, *Chirocephalus diaphanus*, which has the unusual habit of always swimming on its back, propelled through the water by eleven pairs of swimming legs. Even more remarkable is the fact that it is never found in a pond or any other permanent body of water, appearing as if by magic only in temporary pools of water that persist for a short time during rainy periods, living on minute algae and other microscopic organisms that lead a similarly intermittent life.

As soon as they reach full size and maturity the fairy shrimps mate, and the fertilised eggs are retained within the female body in a special brood pouch. When the temporary pools dry up the fairy shrimps die, but the eggs remain alive but dormant until the pool forms again the following year. They then hatch and the cycle begins all over again. The appearance of fairy shrimps in a temporary pool where they have never been found before is probably caused by the transfer of egg pouches in mud adhering to birds' feet. The fairy shrimp is widespread in many parts of the world. It is for example the most common organism in temporary pools formed during the rainy season in Iraq.

The fairy shrimp is a good example of physiological adaptation, for there is good reason to believe that its eggs will not develop until they have been subjected to a period of desiccation, which explains why they are absent from permanent stretches of water.

The most important and numerous of the Branchipoda are the water fleas, which are present, often in incredible numbers, in almost every pond, lake or ditch. The most common kinds are the various species of *Daphnia* and *Simocephalus*. They are most easily recognised by their large branched feathery antennae and by the well developed shell or carapace that covers most of the body.

Water fleas play an important part in the economy of all fresh water in which they live. They feed on the minute algae, which are plants, and are themselves eaten by many kinds of animals, including fish. They thus play a dominant role in the conversion of plant life into animal flesh.

The reproduction of water fleas is most interesting. For most of the spring and summer the whole population consists entirely of females. These produce plenty of eggs, all of which develop without being fertilised, a phenomen known as parthenogenesis. As they are produced they are retained in a brood pouch below the hind end of the carapace, and here after a few days they hatch as miniature replicas of the adult females which laid them. Within a short time they have reached a sufficient size to be able to emerge from the brood pouch to fend for themselves, and a little while longer will see them producing eggs in their turn.

Only as autumn approaches are these parthenogenetic female-producing eggs replaced by a certain number which give rise to males. At least some of the females that continue to appear at this time produce eggs of a different kind, known as winter eggs, which can develop only if they have been fertilised. They are larger than the summer eggs, and contain an abundance of yolk. Each female produces only two or three of these eggs. At the next moult they are enclosed in a special thick-walled container known as an ephippium, which falls to the bottom of the lake or pond, where it remains throughout the winter. The eggs that it contains finally hatch in the following spring to form the first generation of parthenogenetic females.

Recent research suggests that the production of males and winter eggs is probably connected with conditions, rather than being a rhythmical phenomenon occurring after a certain number of asexual generations. So long as conditions are favourable parthenogenesis continues, but lack of food or deteriorating climatic conditions is believed to trigger off sexual reproduction. Normally such adverse conditions occur with the approach of winter. In recent experimental work cultures of various branchiopods have maintained themselves for up to sixteen months under optimum conditions without once producing males or winter eggs. It was found possible, however, to induce them to do so at any time by lowering the temperature, reducing the food supply, or allowing the culture to become overcrowded.

Reproductive habits among the class Ostracoda vary according to the species. In some, both sexes are equally represented and reproduction is normal, whereas in others no males have ever been found, so presumably there is permanent parthenogenesis. In contrast to most crustaceans ostracods do not carry their eggs about with them until they hatch, but deposit them in clusters on aquatic plants. These clusters are fairly easy to find because of their orange colour. There is evidence that ostracod eggs can remain viable in mud even if it dries up, and there are records of ostracod eggs developing normally after having been embedded in dried-up mud for up to thirty years. The nauplius larvae which hatch from the eggs already have tiny bivalve shells.

The importance of the Copepoda both in fresh and sea water has already been mentioned. The copepod body is

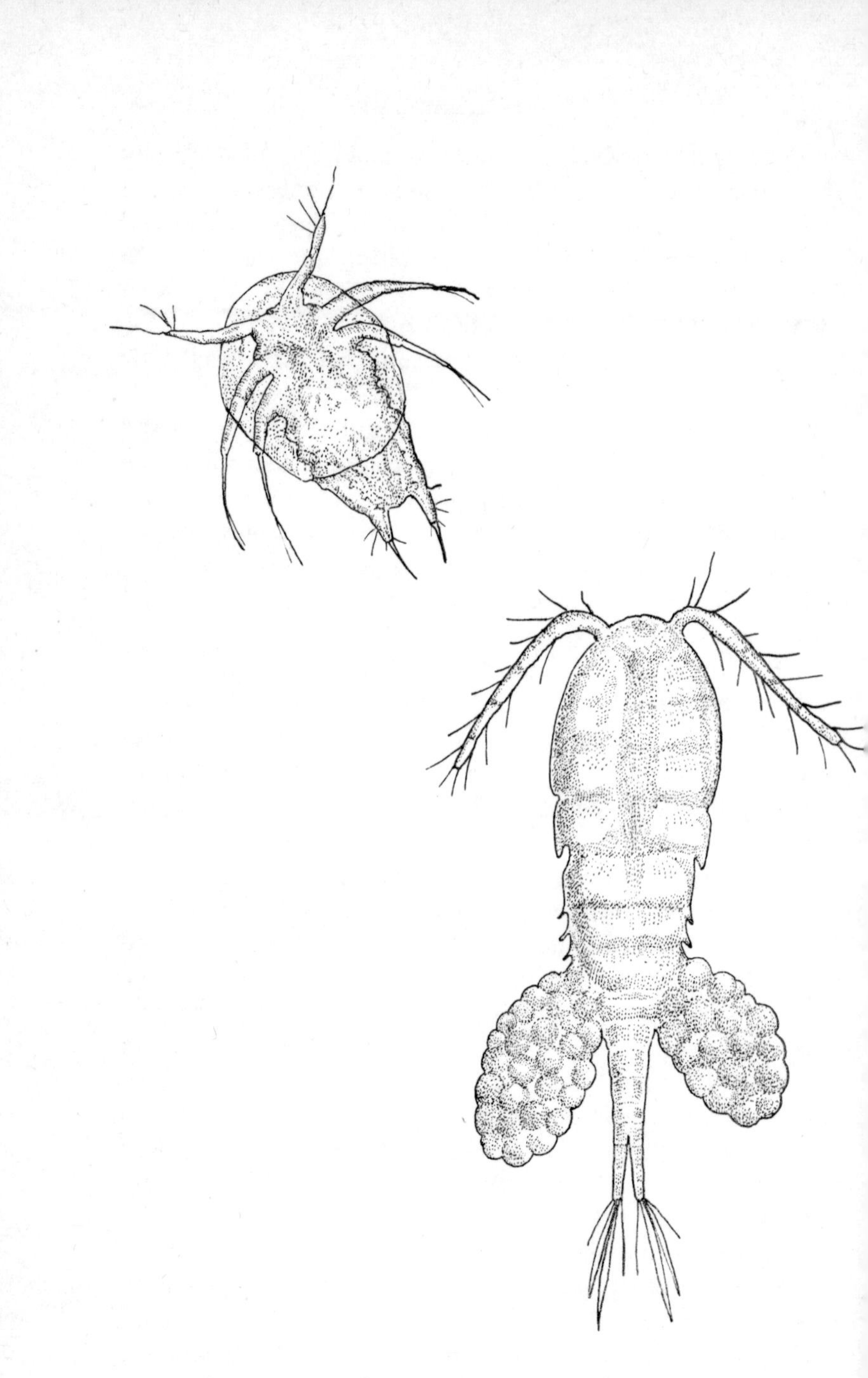

Naplius larva of cyclops, and female cyclops with egg sacs

torpedo-shaped, with a pair of exceptionally long antennules and a pair of well-developed antennae. Another prominent feature is a pair of well developed egg sacs attached to either side of the narrow abdominal region in the female. In some species there is only one egg sac. At the breeding season male copepods transfer their spermatozoa to the females in spermatophores which are placed in the egg duct. Here they remain until the eggs are laid, when they are fertilised as they pass down the duct.

After fertilisation they pass into the egg sacs where they remain until they hatch. In some species two kinds of eggs are produced at different times, one developing immediately while the other is a resting kind, designed to survive during unfavourable periods and develop only when conditions improve.

The copepods go in for larval stages in a big way. Among the most common and abundant freshwater forms are the various species of *Cyclops*, which constitute an important part of the animal plankton of lakes and ponds. *Cyclops* is named after the mythical giant who had a single eye in the middle of his forehead, because it, too, has but a single eye in the middle of its head. In its development it goes through a very complete series of larval stages. It hatches from the egg as a basic crustacean nauplius larva with three pairs of appendages. As they grow these *Cyclops* nauplii moult frequently, and at each moult a little more is added to the hind end, each addition representing one or more of the future adult segments. These larvae with varying numbers of head and thoracic segments behind the original three of the nauplius are known as metanauplii.

The moult that the last metanauplius undergoes involves a drastic change of shape and produces a larva looking much like a miniature adult, but without the full adult number of segments. This is the first copepodid or *Cyclops* larva, and is followed at succeeding moults by four more. The fifth copepodid larva finally moults to produce an adult *Cyclops* complete with all its segments and appendages.

One of the most important of the marine copepods is *Calanus finmarchicus*, which forms the principal food of the herring. Its reproductive habits are very similar to those of *Cyclops*. In Arctic waters it breeds only once each year, but in temperate seas there are usually three or four generations a year, the last one being produced in September. These will have reached the last copepodid stage before further development is inhibited by winter cold, so they do not undergo their final transformation until the following February or March, when they produce the first adults of the year.

A certain number of copepods are modified as fish parasites and their reproduction shows considerable deviation from that of typical members of the group. One of these, *Lernaea*, has an extremely complex life history. The eggs hatch as normal nauplii, which continue development until the first copepodid stage, when they become parasitic on the gills of flatfish, developing a suctorial mouth with which they obtain food from the host tissues. The body now loses much of its crustacean character, entering into what has been described as a pupal stage, in which all power of movement is lost. After a while, however, the ability to swim is regained, and the larvae leave their hosts as adult copepods. Mating now takes place, and is the end of the road for the much smaller males, which do not develop further.

The females, however, do not at this stage lay their eggs, but store the spermatozoa they acquired at mating for future use. They now seek out members of the cod family and attach themselves to the gill filaments of their hosts. Here major degeneration occurs, the head anchoring itself permanently to the host by sending processes deep into its tissues. Only functionless remnants of the thoracic appendages are retained, while the remainder of the body develops into a relatively enormous twisted worm-like sac. Eventually this sac produces great numbers of fertilised eggs. Their development has already been described.

Whereas with most parasites it is the adult stage that is parasitic, in *Monstrilla* the adults of both sexes are free-swimming normal crustaceans, at least in external appearance. The nauplius larvae, however, enter into the bodies of various marine worms, from which they absorb nourishment by means of modified antennae. In this state the larval life is completed, and when a last moult produces the adults, they work their way out of their hosts to live a free life. They are not completely normal, however, for they lack both mouth appendages and gut. All the food they require for their complete life cycle is absorbed during their parasitic larval life.

Closely related to the Copepoda are the members of the fourth crustacean class, the Branchiura or carp-lice. They are all temporary parasites on fish, where they may be found attached to almost any part of the body. In their reproduction they are unusual for crustaceans. The fertilised eggs are not retained in egg sacs, but are deposited on stones, where they remain until they hatch. There are no larval stages, the eggs producing tiny creatures essentially similar in structure and appearance to the adults that produced them.

The barnacles, which constitute the fifth crustacean class, the Cirrepedia, are exceptional in that every member is so different in the adult stage from any normal crustacean that no one would suspect its true relationships without a knowledge of its development.

The most common of all the cirripedes are the seashore acorn barnacles, little cones of shell firmly fixed to almost every rock, and often present in incredible numbers. It has been estimated that there may be more than one thousand million of them along a mile stretch of rocky shore.

For a long time these acorn barnacles puzzled the naturalists, who classed them as molluscs because they looked like tiny limpets. Even today many people think that they are young limpets. A study of their development from the egg, however, showed that they were crustaceans. Unlike most sedentary marine animals, which merely shed their reproductive products into the sea at the breeding season, where they rely upon random fertilisation, the acorn barnacles are able to mate. Each individual is hermaphrodite, but self fertilisation does not occur. Instead, each individual deposits its spermatozoa within the shell of a neighbour by means of an extensible penis.

The eggs are retained within the shell until they hatch as typical nauplius larvae, distinguished from the nauplii of other crustacean groups only by the fact that the front edge is extended laterally into a pair of short horns. After several moults the last nauplius is succeeded by what is known as a cypris larva, on account of its superficial resemblance to the ostracod *Cypris*. It has a pair of tiny hinged shells, six pairs of thoracic appendages and a pair of well developed antennules.

The cypris larva does not feed, for its function in the barnacle life history is to find a suitable spot for the future adult to spend its life, and to fix itself there securely. This settling is no haphazard business. A larva may search the rock surface for upwards of an hour before it finds a suitable position. Having done so, it then stands upon its head, bringing its large antennules in contact with the rock. They, in fact, are connected with a pair of cement glands in the head which now secrete an adhesive liquid which soon dries on contact with the sea water. In this way the larva fixes itself securely to its chosen position.

Further changes now convert it to an adult. The head degenerates completely except for the mouth, and the abdomen is likewise completely obliterated. At the same time the thorax becomes tipped over as it were on to its back, so that the six pairs of now well-developed appendages point upwards. The whole body is now enclosed within a shell consisting of ten plates, six forming the side wall and four the lid. The function of the limbs is to trap minute food organisms suspended in the water which passes through them.

When the cypris larvae first settle they always do so along the flow of the tide, so that they face up or down the shore. Before finally cementing themselves down, however, they turn through ninety degrees, and in doing so ensure that their food net of limbs will always be set across the water flowing on the incoming or outgoing tides. In this way they ensure that they will filter the maximum amount of water and so trap the maximum amount of food.

Around the shores of Britain there are two species of acorn barnacle, *Balanus balanoides,* which is a northern species and *Chthamalus stellatus,* which is a lusitanean or Mediterrarean species. Only on the south-western shores of Britain are both species found, and here *Balanus* is living at the extreme southern limit of its range and *Chthamalus* at its extreme northern limit. This has an interesting effect on their breeding. In the south-west the threshold breeding temperature for *Balanus* occurs in mid-winter, so that by late spring its larvae are ready to settle. The threshold breeding temperature for *Chthamalus,* however, is not attained until midsummer, so its larvae are not ready to settle out of the plankton until the autumn. Further south temperatures would remain too high all the year round to stimulate *Balanus* to breed, while on other British coasts the temperatures necessary to stimulate *Chthamalus* to breed would never be reached.

The cirripede order Rhizocephala consists entirely of parasites which in the adult stage are so degenerate that they bear no resemblance at all to Crustacea, or for that matter to any other group of animals. The best known of these so-called sac barnacles is *Sacculina carcini* which parasitises the common green shore crab *Carcinus maenas.* The external evidence of infection is a yellowish structureless sac wedged between the abdomen or apron and the shell, forcing the apron away from its usual position in close contact with the shell.

The appearance of this external sac is the last stage in the life of the parasite, the actual infection having occurred some considerable time before. In fact it contains the eggs which are passed out into the water through a small opening at the free end to hatch as typical cirripede nauplii with the two characteristic lateral horns at the front end. These nauplii, however, are without mouth or alimentary canal. They eventually give rise to cypris larvae, and it is these that settle on the shore crab. Each cypris larva attaches itself to the base of one of the shore crab's setae by means of one of its antennules. The whole trunk region is now shed, leaving only the very underdeveloped larval head. From the attached antennule a dart-like process is produced, and this penetrates the thin cuticle of the crab at the base of the seta, the remainder of the depleted body following it into the body of the crab. This structureless mass of tissue is now carried round in the crab's blood stream until it reaches and becomes attached to the under-side of the intestine. Rootlets now grow out in all directions until they have penetrated to every part of the crab's body, even to the extremity of its limbs. The food materials necessary for this extensive growth are absorbed from the host's blood. At the hind end of the crab's thorax *Sacculina* develops a sac-like structure consisting essentially of reproductive organs. This presses upon the ventral integument of the crab just at the junction between the thorax and the abdomen. When the crab next moults this sac pushes its way out of the crab's body to become the external sac already noted.

The sixth and last crustacean class, the Malacostraca, contains greater numbers of types than the other five classes put together, and includes all the larger crustaceans. In their development many Malacostra have dispensed with free-living larvae, their eggs hatching as miniature replicas of the adults. Even in those species that do produce larvae the nauplius stage is usually bypassed, the most common first larva being a zoea, a type of larva not found among the other classes.

The class contains three small and relatively unimportant subclasses and two large ones, the Paracarida and the Eucarida. All members of the Paracarida retain their eggs within a brood pouch throughout their development, and in some species even after they have hatched the larvae remain there until they have grown larger, so that when they do finally emerge they are better able to fend for themselves. The best known members of the subclass are the opossum shrimps, the shore slaters, the woodlice, the freshwater shrimps and the sand hoppers. In some species of woodlice a curious phenomenon known as monogeny occurs, in which the members of each brood consist entirely or mainly of the same sex. Why or how this occurs is not known.

A number of smaller kinds live buried in wet sand on the lower parts of the shore, where they feed on debris contained in the sand. There is therefore no necessity for them to emerge each time the tide comes in. One kind, *Bathyporeia*, shows a most interesting rhythm in its appearance. It emerges only at night, and is found swimming in the water in considerable numbers when the tide is in only at fortnightly intervals. It is believed that these intermittent mass appearances may coincide with mating and release of young into the water. The eggs are known to take about a fortnight to develop, so it is possible that for any individual female mating takes place at one appearance and liberation of the young at the next.

The majority of the Eucarida adopt similar methods of reproduction. During the summer months prawns moult every two or three weeks, and one of these moults in the female represents a breeding moult. As with shrimps and lobsters, each abdominal segment carries a pair of paddle-like swimmerets which are in fact used to propel the animals through the water. The female prawn emerges from the breeding moult with long hair-like setae growing on her swimmerets. Before the new shell has had time to harden she mates, and then lays her eggs, attaching them to the setae by means of a cement-like substance produced by special glands on the last pair of swimmerets. A female prawn in berry, as she is described when she has her fertilised eggs, may be carrying as many as two thousand of them. They finally hatch as zoea larvae, and soon after this the female moults again, when she loses the setae that constituted her breeding dress.

Shrimps and lobsters are essentially similar to prawns in their method of reproduction. The shrimp, however, is a sand-burrowing animal, living during the daytime buried just beneath the surface of the sand in the shallow offshore waters, coming up to the surface to feed at night. Just before her eggs are due to hatch the female shrimp travels out to sea to release the larvae.

By virtue of the fact that it lives in fast-moving streams mainly in chalk and limestone districts, the little freshwater crayfish, *Astacus fluviatilis*, has had to modify the typical eucarid method of reproduction. Mating takes place in the late autumn, and begins with the male turning the female over on her back and then shedding his seminal fluid over her abdomen. It is a sticky liquid, and adheres to the hairs on the swimmerets. The female now retires to her burrow, where she lays her hundred or so eggs. These she attaches to the swimmeret hairs, and in this way they come into contact with the seminal fluid and are fertilised. She carries them about with her all through the winter, and in the spring they finally hatch as tiny crayfish. Usually they will remain attached to her swimmerets until they have grown somewhat before breaking away to lead their own lives. This direct production of young without intermediate larval stages represents a necessary adaptation to life in rivers. Free-swimming larvae would be swept down the river and out to sea.

In crabs the powerful tail of the shrimps, prawns and lobsters has become reduced in size and is carried permanently tucked forward beneath the broad cephalothorax. This apron, as it is called, is very narrow and pointed in the male, but quite broad and consisting of seven distinct segments in the female. In the male only two pairs of swimmerets are now represented, and these have become modified as organs for transferring the spermatozoa to the female at the breeding season. But in the female four pairs have survived, and these are well supplied with setae to which the eggs are attached after they have been laid and fertilised.

The crab zoea larva that hatches from the egg is characterised by a long spine which projects backwards from the middle of the dorsal surface of the thorax, and an equally long rostrum which projects forwards. As they grow these zoea larvae undergo several moults without any fundamental change of shape. Then eventually comes a further moult which results in a completely different megalopa larva. It does look more like a crab, with a pair of enormous eyes from which it derives its name. It has a broad flat carapace, five pairs of well developed thoracic limbs and an abdomen carried outstretched and provided with five pairs of tiny feathery appendages. Little change is required at the next moult to convert this larva into a tiny crab.

CHAPTER THIRTEEN

INSECTS: PART ONE

The subphylum Insecta, as already indicated in the previous chapter, is a quite enormous group of animals containing a great variety of types. Before any study of insects can be attempted it is therefore useful to have some idea of their classification.

Altogether the class contains twenty-nine orders divided into two subclasses, the first being the subclass Apterygota, the wingless insects, which are also the most primitive living insects. The four orders in the subclass are believed always to have been wingless, unlike certain other orders whose present-day members are without wings but which were evolved from ancestors that did have wings. The members are all small insects with nothing unusual about their methods of reproduction. The eggs hatch to small individuals which differ from the adults only in their smaller size, and they reach full size through a series of moults, increasing their size slightly at each moult.

To give some idea of the abundance of the different types of insects the approximate number of species at present known is given in brackets after the name of each order. Given the rate at which new insect species are being added to the list, it is impossible to give accurate figures.

The following are the four orders of Apterygota.

Order Thysanura (350). Three-pronged bristle-tails. Includes the silverfish.
Order Diplura (400). Two-pronged bristle-tails.
Order Protura (100). Minute insects seldom exceeding

one millimetre in length.

Order Collembola (2,000). Spring-tails.

The second subclass is the Pterygota, the winged insects. This is divided in turn into two sections, the Palaeoptera and the Neoptera. The former contains only two orders, the Odonata (5,000), consisting of the dragon-flies and damsel-flies, and the Ephemeroptera (1,500), which are the may-flies. All are very dependent upon water, spending the greater part of their lives as aquatic nymphs and only a relatively short time as flying adults. The remaining twenty-three orders are grouped together in the second section, the name Neoptera referring to the fact that all the members have what are described as 'new type' wings, which, unlike those of the dragon-flies and may-flies, can be completely folded back to lie flat along the abdomen.

These last twenty-three orders are again divided into two divisions, the Exopterygota or Hemimetabola and the Endopterygota or Holometabola. The difference between them concerns their method of development. After hatching the Exopterygota develop gradually towards the adult stage, without any sudden drastic change at any stage, whereas the Endopterygota hatch as larvae, which bear virtually no resemblance to the adults, and these larvae are then converted to adults by a drastic reorganisation of the body which takes place during a pupal stage.

The following is a list of the neopteran orders and the kinds of insects they contain.

EXOPTERYGOTA

Order Plecoptera (1,500). Stone-flies

Order Grylloblattodea (5). Small, almost unknown group of insects living on edges of glaciers

Order Orthoptera (10,000). Grasshoppers and crickets

Order Phasmida (700). Stick-insects and leaf-insects

Order Dictyoptera (1,200). Cockroaches and mantids

Order Dermaptera (1,200). Earwigs

Order Isoptera (1,700). Termites
Order Embioptera (150). Small group sometimes known as web-spinners
Order Zoroptera (20). Small group of obscure insects
Order Psocoptera (1,100). Book-lice
Order Mallophaga (2,600). Biting-lice or bird-lice
Order Siphunculata (Anoplura) (250). True lice or sucking-lice
Order Hemiptera (55,000). The largest and most diverse Exopterygota order. General name, bugs. Includes the bed-bug, cicadas, lantern-flies, frog-hoppers, aphids, scale-insects, mealy-bugs, water-bugs, water-boatmen, water-scorpions and pond-skaters
Order Thysanoptera (2,000). Thrips

ENDOPTERYGOTA

Order Neuroptera (5,000). Alder-flies, snake-flies, lace-wings and ant-lions
Order Mecoptera (350). Scorpion-flies
Order Trichoptera (4,500). Caddis-flies
Order Lepidoptera (200,000). Butterflies and moths
Order Diptera (85,000). True flies
Order Siphonaptera (1,000). Fleas
Order Hymenoptera (100,000). Ants, bees, wasps, saw-flies, wood-wasps, chalcids, gall-wasps, ichneumon-flies
Order Coleoptera (275,000). Beetles and weevils
Order Strepsiptera (300). Females mostly wingless, many parasitic on other insects

Growth in insects as in other arthropods involves a series of moults or ecdyses, because little or no growth is possible once the cuticle has hardened. Ecdysis takes place in two stages. In the first stage the insect detaches its body from the old cuticle and lays down a new cuticle inside it. The second step involves splitting the old cuticle and breaking out of it. It is this latter step that is usually referred to as moulting. The first stage, because it takes place within the old cuticle, cannot be observed. Usually there is only a very short interval between the two stages, but in some cases a considerable time elapses after the insect has entered the next phase before it breaks out of the old cuticle. This is known as a pharate or concealed stage. It occurs most commonly in species that survive the winter as eggs. In some the development of the fertilised egg slows down or ceases, to be resumed in the spring. But in others development proceeds at the normal pace within the egg shell until a fully developed embryo ready to hatch has been produced. This then remains within the shell until the spring as a pharate nymph or larva, depending upon the type of insect.

Ecdysis occurs only during the development stages in insects, except in the case of the four orders of the primitive wingless insects, the Apterygota, in which moulting and growth can continue throughout life. The stage between any two ecdyses is referred to as an instar.

The great difference between the Exopterygota and the Endopterygota is in their development. The Exopterygota hatch as nymphs which in many cases bear quite a resemblance to their parents, and each instar progresses a little further towards the adult form, the last moult giving rise to the complete adult. The total number of instars between hatching and the adult varies very considerably between different orders. In many it varies between five and eight, but in those insects that have aquatic nymphs there may be as many as forty moults before the adult is finally produced. These aquatic nymphs are sometimes referred to as naiads.

In the Endopterygota the larvae that hatch from the egg bear no resemblance to the adult, and after a number of larval instars there is an abrupt change in development when the larva changes to a pupa. It is in this pupal stage that the completely different larva is converted by a drastic internal reorganisation into the final adult or imago.

There are in fact three types of larvae in the Endopterygota. The most active kind have a well developed head and thorax and are usually carnivorous, feeding on small animal life.

They are known as campodeiform larvae, because they bear a marked resemblance to the Thysanuran bristle-tail *Campodea*. Much more common are the less active eruciform larvae, such as the caterpillars of butterflies and moths, which feed on plants and have enormous appetites. They have three pairs of short legs on their three thoracic segments and may also have a few pairs of pad-like prolegs or false legs on some of their abdominal segments, whose function is to support their well filled abdomens. The third type of larva is the grub, devoid of all appendages and capable only of a wriggling movement. These are the typical larvae of flies, and of ants, bees and wasps.

But whatever the form of the larva it eventually reaches the time when it changes abruptly to the last pre-adult instar, the pupa. Despite its apparent immobility, the pupa is anything but a resting stage. True, it does not feed, and in the majority of cases does not exhibit any external activity. Within, however, there is great activity, because the larval body is undergoing a drastic reorganisation to convert it into a fully formed adult. In some insects the complete pupa is contained within a cocoon, often constructed of silk produced by the last larval instar, though in some cases made from extraneous materials such as sand or earth. In many flies the cocoon consists of the last larval cuticle which is not shed when the last larval instar changes to a pupa.

In the remainder of this chapter and the two following chapters we shall examine the various reproductive methods found in the more important insect orders.

Adult dragon-flies and damsel-flies have a comparatively short life of about four weeks, with a nymphal life extending for up to two or three years. Their method of mating is quite unlike that of any other insects. The male reproductive organs are situated on the last but one abdominal segment, opening on the ventral surface. But before mating occurs the male bends his body forward and transfers the spermatozoa to special pairing organs situated on the second abdominal segment.

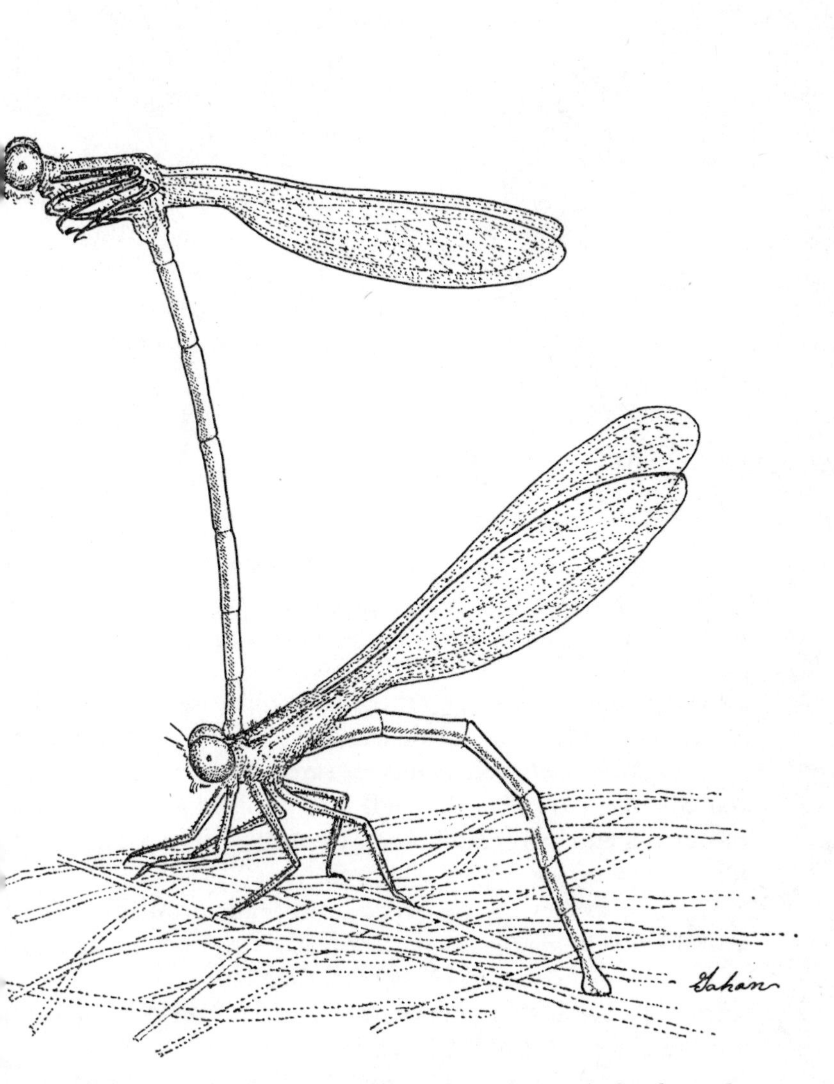

Damsel-flies laying eggs. The male holds the female firmly as she deposits her eggs on water plant fibres

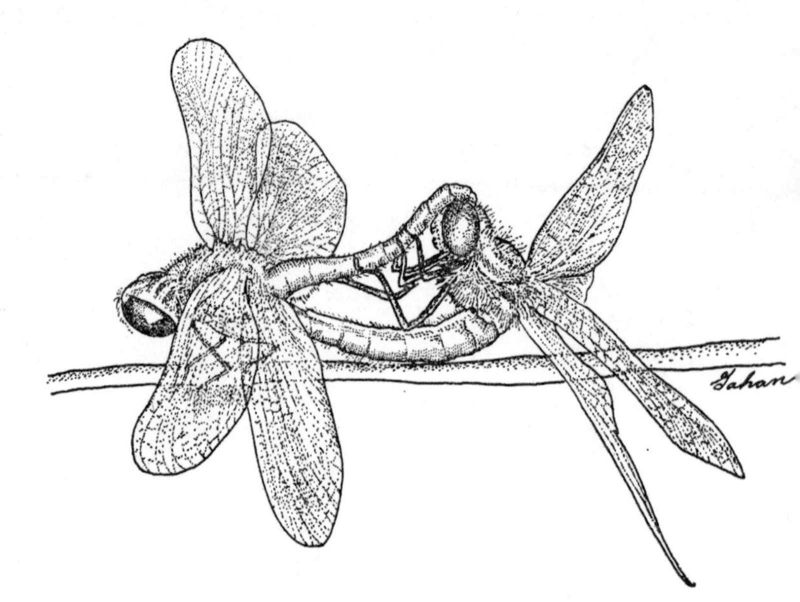

Mating dragon-flies (female has tip of abdomen brought around under male's abdomen where his sperm capsules are stored)

Having achieved this transfer the male then seizes a female by her head with his tail appendages and flies with her to rest on a plant. Here the female bends forward her abdomen until the reproductive organs on her tail end come into contact with the male's pairing organs, when transfer of spermatozoa and fertilisation of the eggs is achieved.

In some species the pair remain in tandem until the eggs are laid, first flying to an aquatic plant where they rest an inch or two above the water. The female then partly or completely submerges her body to deposit her fertilised eggs on the submerged parts of the plant. Sometimes she even makes incisions in the stem of the plant into which the eggs can be inserted. But the majority of species merely drop their eggs into the water as they fly over it, sometimes dipping the tips of their abdomens into the water to wash off the eggs.

Dragon-fly nymphs are plump-bodied, while those of the damsel-flies are slender and stick-like in appearance. But both, like the adults, are entirely predatory. They have an extremely efficient structure with which they capture their living prey. This is a specially modified labium, carried normally folded beneath the head, but being capable of being shot forward whenever suitable prey approaches. At its end it bears two pincer-like structures with which the prey can be impaled. These larvae are among the most ferocious of all the small animal life found in fresh water. Normally they feed on small animal life, but the larger nymphs will capture worms, tadpoles and even small fish.

The total number of instars between the hatching of the egg and the emergence of the fully formed imago varies. The smaller species complete their development in one year, but the larger species may take two or three years, and incorporate as many as twelve instars.

Dragon-flies are among the oldest of our existing insects. In the days when the coal measures were laid down, some three hundred million years ago, enormous dragon-flies with wing spans of up to two feet were common, sufficient evidence in itself that there must also have been an abundance of smaller insects on which they could feed. These early dragon-flies were in fact the largest insects that have ever lived. Like the dinosaurs, their very size must have been a fatal evolutionary handicap to them, for they eventually gave rise to smaller forms which in turn gave rise to the highly successful modern descendants.

The Ephemeroptera or may-flies, like the dragon-flies, spend the majority of their lives as aquatic nymphs. In contrast, however, the nymphs are vegetarian and the adults do not feed at all. Their sole function is to mate and lay their eggs, and for this they need only a short period of time. In many species the adults live for only a day, and in none of them is adult life prolonged for more than two or three days. The name Ephemeroptera means 'winged insects which live for a day'.

Adult may-flies are not easily confused with any other adult insects. They have two or three enormously long tail filaments and a very large pair of forewings, the hind pair being much smaller. Their flight is very weak, and they remain over the water from which they have emerged. The males fly in swarms above the water, the swarm itself remaining more or less stationary while the individuals within the swarm indulge in a kind of vertical dancing flight, fluttering upwards a foot or two with rapid but rather weak wing beats and then falling back to their starting points with motionless wings.

These swarm dances of the males seem to play an important part in mating. If a female approaches the male swarm, several of the males will detach themselves in pursuit, the first to reach the female mating with her at once. It is probable that the males are attracted by some kind of scent given off by the female.

In many species the mated female turns her attention to egg-laying immediately after she has mated, while in others egg-laying may be delayed for a few hours or even a few days. Whenever it occurs egg-laying is followed almost immediately by the exhausted female dropping on to the surface of the water. Many of them will be snapped up by fish.

In most species the eggs are washed off the tip of the abdomen as they are laid, the female frequently dipping her abdomen into the surface of the water for this purpose. Some vary this procedure by dropping their eggs on to the water from some distance above the surface. A minority of species adopt a more precise method of depositing their eggs. They land on some part of an aquatic plant projecting above the surface of the water and then crawl down it until they are below the surface, where they deposit their eggs on the submerged parts of the plant.

Some species live in still or slow-moving waters, while others are only found in fast- or medium-flowing streams. Some, too, are sluggish while others are quite active. And the duration of nymphal life varies from six months to two or three years. Among those species that have a long nymphal life there may be as many as thirty nymphal instars. But however long the nymphal life and the number of instars the may-flies show one feature that is unique among insects. The last nymphal instar possesses functional wings, and is known as a subimago. These subimagos come out of the water, and eventually moult for the last time to produce the true imago. To the fisherman the subimagos, which are rather dull in colour, are known as duns, while the final imagos, which are much more brightly coloured, are known as spinners.

The members of the first order of the Exopterygota, the Plecoptera or stone-flies, also have aquatic nymphs. The adults, too, have only a short existence, varying between a few days and three weeks, following a nymphal existence which occupies at least a year. The most important function for the adults to carry out during their short life is of course mating and egg-laying. After mating above the surface of the water from which they have emerged the females of the larger species usually swim on the surface of the water during egg-laying, so that the eggs pass straight into the water. Females of smaller species usually fly just above the water surface, dipping their abdomens momentarily into the water each time an egg is extruded so that it can be washed off. The eggs sink quickly to the bottom, where they become firmly stuck to stones. In some species they are provided with an adhesive disc at one end, by which they become anchored, while in others attachment is effected by a complete coating of adhesive jelly. Adhesion to the river bed is necessary in fast-moving streams or the eggs would be swept away.

Stone-fly eggs usually hatch in about three to five weeks, though in some of the larger species the period of incubation may be as long as three months. Most species are found only in swift mountain streams where the water is cold but well oxygenated, and the ground is clean and stony without mud or silt deposits, and often without much weed cover. So adapted are they to this kind of water that warmer water kills them, and they cannot live in aquarium tanks because the water contains too little oxygen.

The nymphs can crawl actively on their quite powerful legs but they do not swim. Fishermen know them as 'creepers', and they are frequently used as bait in mountain streams. The duration of nymphal life seems to depend upon the size of the species. The smaller kinds live as nymphs for one year before changing to adults, whereas the larger kinds may take as long as three years to complete their development, and may moult thirty or more times in the process. Nymphs of the smaller kinds seem to be entirely vegetarian, but the larger kinds may be carnivorous, feeding on the nymphs of may-flies and other stone-flies, and the aquatic larvae of caddis-flies and *Chironomus* midges.

When the time comes for the nymphs to emerge from the water and undergo the final moult by which they are transformed into adults they climb up the stems of water plants or up the river bank. Having found a suitable site for their transformation they emerge from their last nymphal skins through a slit that appears down the back of the head and thorax. The folded wings are blown up by body fluid being pumped into their veins under pressure. Then follows a resting period during which they dry and harden. After this the adult stone-fly is ready to take to the wing for its final brief existence.

Stone-flies are as popular as may-flies among fly-fishermen, both the adults and the nymphs being used. The larger stone-flies have an historical claim to fame as the earliest known flies to have been artificially copied by anglers. Nearly five hundred years ago a book called the *Treatise on Fishing with an Angle* was published, and contained detailed instructions for making up imitations of the larger stone-flies.

The members of the next three orders used to be classed together in the single order Orthoptera, but modern classification includes only the grasshoppers, locusts, bush-crickets and crickets in this order. The stick-insects and the leaf-insects form the new order Phasmida, and the order Dictyoptera comprises the cockroaches and mantids. The order Orthoptera can be divided into three super-families: the Acridioidea, comprising the true or short-horned grasshoppers and locusts; the Tettigonioidea, the long-horned grasshoppers, known in North

America as the katydids (referring to the sound that they make); and the Grylloidea or crickets.

The true grasshoppers lay their eggs in batches, each batch consisting of a dozen or so eggs enclosed in a tough case known as an egg-pod. After the eggs have been laid they are covered with a frothy liquid which hardens to a spongy consistency and provides a protection for the eggs during the autumn and winter. In most species the egg-pods are deposited in the soil, but some deposit them at the base of grass stems. They are produced during the summer months, and then remain dormant throughout the autumn and winter. Not until the following April will the eggs begin to hatch, but they do not immediately produce the first nymphal instar. They emerge as worm-like creatures still enclosed in a transparent covering. This is soon shed to reveal the first instar. In most species four nymphal instars separate the vermiform larvae from the adult imagos, but in some species there may be more than this number. Adults do not begin to appear until the latter part of June, and are not common until July.

The long-horned grasshoppers are more familiarly but confusingly known as bush-crickets. In contrast to the true grasshoppers they lay their eggs singly, some species in the ground, some in the crevices in bark, and the remainder in slits which they make themselves in plant stems. Again in contrast to the true grasshoppers, which have short ovipositors, the female bush-cricket has a long ovipositor which is sharp and flattened, and in some species serrated at the tip for use as a saw.

After the eggs have been laid they undergo a resting period which carries them through the winter. They seem to need rather higher temperatures than the eggs of grasshoppers need before they will hatch, because on the average they hatch up to six weeks later, in May and June for most species, the adults maturing in July and August.

Again like the grasshoppers, bush-cricket eggs hatch as vermiform larvae which soon shed their larval skins to emerge as the first nymphal instar. Most species go through five or six instars before the imago emerges, but some species, including the great green bush-cricket, have up to nine nymphal instars.

Many of the true crickets have a life history extending over a longer period than that of other Orthoptera. The number of nymphal instars varies between eight and thirteen, and the first instar is preceded by a vermiform larva. The considerable variations between the life histories of different species make it impossible to generalise.

The wood-cricket takes two full years to complete its life history. Eggs are laid in late summer and early autumn and pass through the winter in a resting phase which lasts until they begin to hatch in June. During the remainder of the summer the nymphs usually succeed in reaching the fifth instar, occasionally the sixth. Now follows a resting phase which takes them through the second winter. These resting instars, however, are not undergoing true hibernation, because during mild spells, even in the depths of winter, they may become active. But they never moult. Development is resumed in the following April, when a moult produces the sixth instar. Three more moults follow to give rise to the adult after the eighth nymphal instar. The timetable for these final stages depends upon the spring weather. Under favourable conditions adults may appear in June, but it is possible for last-instar nymphs to be still around in August. Because two full years are required for complete development the adults of one year can never be in a position to mate with the adults of the preceding or the following years. There are thus two distinct strains, usually referred to as even-year and odd-year wood-crickets.

The field-cricket, although it completes its life cycle in a period of twelve months, does undergo true hibernation through the winter. Each mature female produces several hundred eggs between May and July, and these are deposited singly in the ground. After hatching instars succeed each other with considerable rapidity, so that by the autumn the tenth instar has been reached. The first nine instars live entirely above ground, but this tenth one proceeds to dig a burrow in which it will hibernate until the following March. Soon after this it will moult to produce the eleventh and final nymph, and this in turn will moult to produce the adult in late April or early May, depending upon weather conditions.

The house-cricket probably came originally from North Africa or South-West Asia, where it still exists as an outdoor species. In houses and other buildings its life cycle varies considerably in duration according to the temperature conditions. The number of instars seems to vary considerably between seven and thirteen, though in European colonies eleven seems to be the most frequent number. In indoor colonies breeding may take place at any time, so that all stages may be found whatever the time of year.

The mole-crickets are the most modified in structure of all the Orthoptera, the front pair of legs being enlarged and adapted for digging. They spend most of their time in the burrows that they excavate, but during warm periods they may come up to the surface at night. Breeding occurs over a period of several months beginning toward the end of April, the eggs being laid in specially excavated nest chambers which may be up to a foot below the ground surface. Several hundred of them will be produced over a period of up to fourteen days, and they are piled on the floor of the chamber. The female guards the nest and periodically licks the pile of eggs. Her saliva is thought to kill any mould spores that may settle on them. Certainly they usually become mouldy if they are taken away from the mother.

The eggs hatch in two to four weeks, the number of nymphal stages achieved before the onset of winter depending upon the time the eggs were laid. Whatever the stage, the winter will be spent in hibernation at the bottom of burrows which may be as much as three feet below the surface. The total number of nymphal instars is not known for certain, but is probably ten. Those that hatched early in the season are able to complete their development late in their second summer, but will not become sexually mature until early in their third summer. Those that hatched later in the year have probably only reached the later nymphal instars by the time it is necessary for them to go into hibernation again, so they spend two winters as nymphs, and finally develop into breeding adults late in their third summer.

The stick-insects and leaf-insects comprising the order Phasmida are very sluggish creatures, relying upon their well developed camouflage to protect them from their enemies. Parthenogensis is more common among the stick-insects than sexual reproduction. Indeed, in some species males are extremely rare. It has been suggested that the almost universal occurence of parthenogensis may be an adaptation to their extremely sluggish mode of life. During the daytime there is scarcely any movement, but to find a mate they would have to move about, and whereas a stick-insect is extremely well camouflaged so long as it stays put, it becomes very conspicuous when it begins to move.

Stick-insect eggs have thick shells and are usually provided with detachable lids through which the young nymphs can emerge. The eggs are merely dropped to the ground, where they remain for several months, even a year or more, before eventually hatching. The best known species is the so-called laboratory stick-insect, *Carausius morosus*, which is often kept as a pet. It is extremely prolific, each female producing several hundred eggs at the rate of two or three a day, and these take between four and six months to develop unless they are kept in a very warm room, when it is possible for them to hatch in a couple of months. The six nymphal instars that follow take a further four to seven months to complete.

In the cockroaches the fertilised eggs are enclosed in a case known as an ootheca. In some species this is deposited after it is formed; in others it retains its position at the opening of the female reproductive duct; while in a few species it is withdrawn into a special part of the female duct known as the brood sac. When the eggs are ready to hatch those that have been deposited or retained at the entrance to the egg duct emerge through a slit in the ootheca, which opens along a keel, and the nymphs emerge as vermiform larvae. The oothecae that are retained in a brood sac often split open soon after their withdrawal, so that the remainder of the development of the eggs takes place within the brood sac. Whichever method is adopted, the vermiform larvae shed their skins almost as soon as they are exposed to the air to reveal the first nymphal instar. The average number of nymphal instars seems to be six.

The mantids, although superficially very different in appearance, are included in the order Dictyoptera with the cockroaches partly because of their similar method of reproduction. They, too, produce their eggs in batches enclosed in an ootheca or egg-pod. They are notorious for their cannabalism. Larger species will eat smaller members of their own kind, and the females of some of the larger species will proceed to eat the males, starting with the head, while they are in the very act of fertilising their eggs.

At one time the earwigs, comprising the order Dermaptera, were also included in the order Orthoptera. During the winter they go into hibernation, usually in pairs, either in a cavity which they excavate in the soil, or beneath a log or a stone. Sometime during this hibernation mating takes place, and the female lays her eggs in February or March. When they hatch as tiny replicas of their parents she continues to take care of them while they grow, a most unusual feature in more primitive insects, in which parents and offspring generally have no contact with each other. After a number of moults, during which they increase in size and acquire increased colouring, they are at last able to go off to fend for themselves. Because of this maternal protection the earwig does not have to produce so many eggs as other insects whose eggs and young stages are more vulnerable to enemy attack. The number of eggs produced by a female earwig varies from about twenty to about ninety, depending upon the species. Real maternal care, as distinct from the communal protection offered by the social insects, is a rare phenomenon among insects.

The order Isoptera—the termites—are unique among the Exopterygota in having evolved a very elaborate social life. The fact that they are commonly referred to as white ants gives the completely false impression that they are related to the ants, which of course are also highly organised socially. But the ants, bees and wasps, which belong to the order Hymenoptera, are among the most advanced of the Endopterygota, and are therefore at the peak of insect evolution, whereas the Isoptera are comparatively primitive insects. The two orders must have evolved their social systems quite independently. The parallels between the two kinds of insect colonies are more remarkable than the differences.

Like the social Hymenoptera, the termites exhibit polymorphism, each species containing a number of structurally different individuals, each type having its own essential function in the community. But unlike the Hymenoptera, each type or caste, as it is called, contains individuals of both sexes. Reproduction is the function of full-winged males and females, sometimes referred to as kings and queens. Each colony normally contains one active royal couple. In addition to these there are other sexually active short-winged reproductives which are capable of laying fertilised eggs if anything should happen to the royal pair. The majority of the eggs eventually gives rise to workers which, although they are sterile, are of both sexes. Then there are the soldiers, also of both sexes and also sterile. Of these there are two kinds, the mandibulate soldiers, which have greatly developed mandibles for dealing with any enemies which may penetrate the colony, and the nasute soldiers, in which the jaws are insignificant but the top of the head is produced into a long nose-like process at the tip of which is a gland capable of producing a sticky secretion. This can be used to 'gum up' an enemy, which is most likely to be an ant.

As with the Hymenoptera, new colonies begin with a swarming of the full-winged males and females which pair up on a nuptial flight. In some species they mate during this flight, but in many cases actual mating is delayed until they have landed and broken off their wings, and even until they have chosen the site for their colony and have begun to excavate the nest. When the first eggs have hatched the royal couple feed the nymphs at first on their saliva and subsequently on regurgitated food. Again unlike the Hymenoptera, in which the larvae are helpless and have to be looked after until they pupate, the termite nymphs are active and can take their place in the economy of the colony from their first instar. As soon as they are able the first few nymphs in the new colony take over the task of looking after the royal couple, who from now on can devote all their energies to producing fertilised eggs. In many species the queen grows bigger and bigger, until she is little more than an enormous bag producing eggs, sometimes in incredible quantities.

There are two main types of termites, a more primitive type, which lives in wood and is responsible in the tropics for the sudden collapse of wooden buildings, and the subterranean termites, which nest in the ground. These are the ones that build the spectacular termite hills that are such a feature in Africa and in other parts of the world where they are found. The queens of the wood termites are not much enlarged and produce only two or three hundred eggs in a year. But the queens of the subterranean species are capable of producing tens of thousands of eggs each season.

The remaining five orders of Exopterygota are thought to have evolved from a separate stem from the other orders in the group. Except for the first two orders the mouth parts have become modified for piercing plant or animal tissues in order to suck up their juices. Little is known about the first order, the Psocoptera, which includes the tiny book-lice. Whether they really do eat paper and bindings, or merely feed on the moulds that have already attacked them is not known for certain.

The members of the order Mallophaga are usually known as biting-lice or bird-lice, because they feed on skin debris and are mostly found on birds, though there are a few species which live on mammals. They are all parasites. Their eggs are laid singly and are attached to the feathers. The young instars that emerge from them resemble their parents both in structure and habits, and in a few weeks have become adult in their turn, having moulted only two or three times.

The true lice or sucking-lice forming the order Siphunculata are all mammal parasites, almost every species of mammal having its own exclusive louse. There is even an elephant louse! They lay large numbers of eggs, and there are only three nymphal instars, so that the whole life cycle takes only about a month. The human louse is the vector for the transmission of typhus and trench fever.

Sucking reaches its peak in the Exopterygota in the bugs, a varied collection of insects forming the large order Hemiptera. With its fifty-five thousand species, vastly more than are found in any other order of the Exopterygota, the Hemiptera must be recognised as easily the most successful order in the group, rivalled only by the four most successful orders of the Endopterygota. The order is divided into two suborders, the Homoptera and the Heteroptera. All the Hemiptera have mouth parts modified for piercing plant tissues or other animals in order to feed on their juices.

The best known members of the Homoptera are the aphids —the green-fly and the black-fly. To most people green-fly are associated with roses, and there is always something mysterious about their sudden appearance just after the new shoots have started to grow in the spring. One weekend there is not a green-fly in sight, but by the next the young shoots can be quite smothered with them.

To understand where these have come from we must go back to the previous autumn. During late September winged males and females are produced, and these leave the roses and fly to the trunks of nearby trees. Here the females produce a generation of wingless females by parthenogenesis, and it is with these females that the winged males mate, an event that is followed by the females laying rather large eggs with thick black protective shells. These eggs lie dormant in the crevices on the bark of the tree right through the winter, able to withstand the most severe weather unharmed.

With the onset of warmer weather the following spring these eggs finally hatch, each one producing a wingless female. Not a single male is produced. These stem mothers, as they are called, reach maturity in about ten days and then reproduce in their turn by parthenogenesis. Some of these offspring are winged, the others wingless. The winged aphids fly away from the winter tree and on to the roses, which by now are shooting well. Here they reproduce in their turn, once more by parthenogenesis, so building up the colony on the rose bush. The wingless females left behind on the winter tree go on reproducing, so providing a reservoir from which other roses, and indeed all other plants on which aphids are found, can be colonised.

From now on, right through the summer, generation follows generation in rapid succession, all produced by parthenogenesis, with a regular alternation between winged and wingless generations. The wingless aphids serve to build up the population on their own plant as it grows in size, while those with wings provide an additional reservoir from which new plants can be colonised. Only in the autumn is the single generation of males produced to mate with the last generation of wingless females, as we have just seen. Among the aphids the role of the male has indeed been reduced to a minimum.

The majority of the Homoptera are small soft-bodied insects like the aphids, but there are a few larger kinds, notably the cicadas. Because of the incessant sound they make on hot sunny days many people believe them to be related to the equally noisy grasshoppers and crickets. In fact of course they are not at all related, and they produce their sounds by an entirely different method. In general they prefer warm climates.

One North American species, which is commonly known as the harvest-fly, has a longer larval life than any other known insect. In the southern states the adult cicadas emerge thirteen years after the eggs were laid, but in the rest of North America seventeen years elapse between one generation and the next. Appropriately this species is named *Cicada septemdecim*. The whole of this protracted larval life takes place underground.

When the adults do finally emerge they live for a month or so, during which time of course they mate and lay the eggs that will result in the next generation in seventeen years time. Until the life history of these cicadas was known it was a complete mystery why in a certain district there would be a sudden swarming of these noisy insects lasting for a few weeks, to be followed by many years when no cicadas appeared at all. Of course the colonies are not all synchronised throughout North America, so that in any year some districts will have their cicadas.

The suborder Heteroptera contains two groups, one consisting of a variety of aquatic insects, including the pond-skaters, water-gnats, water-scorpions and water-boatmen, the other consisting of the capsid-bugs, shield-bugs and the notorious bed-bug. Of the aquatic members some live on the surface of the water and the others within it.

Of all the insects capable of walking on water the most familiar are the pond-skaters, sometimes known as water-striders. The secret of their ability to walk on water is the fact that the legs, feet and under-side of the body are covered with a pile of fine hairs which prevents their becoming wetted. Their eggs are laid in clusters enclosed in a mass of jelly and attached to aquatic plants beneath the water surface. The much more slender and slower-moving water-gnats lay their eggs singly on plants growing at the water's edge, while the water-crickets lay theirs on floating vegetation. All these water walkers live by sucking the juices from dead or dying insects that fall on to the water. Water-scorpions and water-boatmen, which live beneath the surface, lay their eggs singly in slits that the female makes in the stems of plants.

The last of the hemimetabolous Exopterygota are the thrips, comprising the order Thysanoptera. These are very tiny insects which suck plant juices and are particularly abundant in the flower heads of the Compositae, the daisy family. Their wings have a most peculiar structure. Each true wing is reduced to a narrow strap, and this is fringed along both the front and hind edges with long hairs, giving the wing a feather-like appearance.

Their most interesting feature, however, is their life history. There are two suborders, the Terebrantia, which have a saw-like ovipositor with which they deposit their eggs within the tissues of plants, and the Tubulifera, whose ovipositor is a simple tube, sufficient to enable the females to deposit their eggs in crevices. In the Terebrantia there are four nymphal instars, and five in the Tubulifera. In both the first two instars are similar in appearance to the adults, as in all the Exopterygota. They are active nymphs, moving about and feeding. The two or three following instars however are inactive and do not feed, and in some species they are actually enclosed in a cocoon. They thus resemble the pupal stage in holometabolous insects. The last instar is in fact often referred to as the 'pupa', the one preceding it being called the 'prepupa'.

CHAPTER FOURTEEN

INSECTS: PART TWO

In this chapter we turn to the holometabolous Endopterygota. There are four extremely successful orders, which account between them for more than six hundred and fifty thousand of all the known insect species, and five other orders which contain between them just over eleven thousand species. But it is three of these orders representing the more primitive Endopterygota that we consider first in this chapter.

The main function of holometabolous larvae is feeding, and in many cases they have accumulated sufficient material by the time they pupate that the adult has no need to feed at all, and can concentrate on its main function of reproduction. Because, too, they are so different in structure the larval and adult stages of these insects often lead completely different lives and have different feeding habits in those species in which the adults also feed.

Among the Endopterygota three main types of terrestrial larva are recognised. The most active kind are known as campodeiform larvae. Their three pairs of thoracic legs are well developed, and the cuticle of the head and thorax forms a thick protective covering. They are typically voracious carnivores, feeding on other insects and small animals. The eruciform larva is less active, and its thoracic legs are not so well developed. The caterpillars of moths and butterflies are typical examples. They tend to be plant feeders, eating enormous quantities of food and developing fat flabby bodies which are only capable of cumbersome crawling movements. Finally there are the completely legless apodous larvae usually referred to as grubs. These are the typical larvae of bees, wasps and flies. The fourth type of larva found among the Endopterygota are the various kinds of aquatic larvae.

The larvae of all the Neuroptera are carnivorous campodeiform larvae, some having mouth parts designed for chewing, others for sucking. Adult alder-flies are quite common during the summer among vegetation growing beside fresh water. They lay their eggs several hundred at a time during May and June, attaching them to the leaves of grasses and sedges. In about two weeks the larvae hatch and make their way to the water, where they bury themselves in the mud at the bottom, only coming out to feed. They are provided with a pair of strong pointed mandibles with which they both capture and cut up their prey, which consists of other insect nymphs and larvae, small crustacea and worms. The larval life extends over a full two years, at the end of which time the last larval instars leave the water, bury themselves in the ground and pupate. In about a fortnight the adult insects emerge. Although they, too, possess powerful jaws it is believed that in many species at least they do not feed.

The snake-flies are very similar in appearance to the alder-flies, with thin gauzy wings that cover the sides of the body like the two sides of a roof. Their distinguishing feature is a long 'neck' formed by an extremely elongated first thoracic segment holding the head well forward from the remainder of the body. Their larvae, however, are completely terrestrial, running about beneath loose bark on trees in search of other insects on which they feed, cutting them up with strong mandibles similar to those of the alder-flies.

In the remainder of the Neuroptera the mandibles are modified for sucking the juices out of their victims instead of cutting them up. Best known of them are the pretty green lace-wings. With their pale green bodies and wing veins and their prominent golden eyes they are common summer garden insects. Their larvae are of considerable value to the gardener because they feed mainly on aphids, each consuming one or two hundred during its lifetime.

The eggs are unusual in that they are attached to leaves suspended at the end of long thin stalks. Before she lays each egg the female exudes a tiny drop of sticky liquid, places it on the leaf, and then raises the tip of her abdomen so as to draw the liquid out into a fine thread. This solidifies rapidly to form the stalk on which an egg is then laid.

The larvae of some species indulge in an unusual method of camouflage, covering their backs with debris such as the dead skins of their victims, pieces of dried leaves and similar materials. Each piece is actually picked up in the insect's mandibles and placed on its back, and is finally settled into position by wriggling movements of the body. Small hooked bristles hold the material in place.

The brown lace-wings are smaller and much duller in appearance than the green species, and therefore less conspicuous. Their habits, however, are similar: they and their larvae feed mainly upon aphids. Their eggs, though, are not laid on the ends of stalks, and they do not indulge in the habit of camouflaging themselves. A few species have quite a different mode of life. They live near to streams and lay their eggs on damp moss at the water's edge. The larvae are semi-aquatic, living among the moss and feeding by sucking the blood of various insect larvae that also live there.

Closely related to the brown lace-wings is a small group of sponge-flies. The adults live by the banks of canals, rivers and lakes. Their eggs are laid in the water, and the larvae that hatch from them are fully aquatic, bearing seven pairs of gills along the abdomen. They spend their lives as parasites of freshwater sponges, living on the surface of the sponge.

Also included in the Neuroptera are the ingenious ant-lions. The adults are similar in appearance to the other members of the order, but the larvae are modified in structure and behaviour for their very unusual method of capturing their prey. They have large flat heads and a pair of long, forwardly directed toothed sickle-shaped mandibles, and they live at the bottom of pits in sand, buried so that only the tips of the mandibles and the antennae are showing.

An ant-lion may wait at the bottom of its pit-fall for hours or even days, but sooner or later a passing ant or other pedestrian insect will approach the rim of the pit and fall in. Immediately the ant-lion will emerge from hiding and pounce on it and give it a bite with its powerful mandibles. The saliva that it injects both kills the victim by poisoning it and reduces its soft internal organs to a nutritive fluid, which can be sucked up by the predator at leisure. Occasionally the victim may succeed in dodging the first attack and begin to climb out of the pit-fall, but it stands little chance of escaping because as it climbs so the ant-lion bombards it with grains of sand which cause it to fall down to the bottom again, where the ant-lion is waiting to pounce on it.

The ant-lions actually excavate the pits themselves by shovelling sand on to their large flat heads with their front pair of legs. As soon as a load has been assembled the head is jerked upwards and outwards, throwing the sand outside the limits of the future pit-fall.

Because food supplies tend to be rather intermittent, the larval stage stakes two or three years to complete, but eventually the last larval instar buries itself in the bottom of its pit and pupates within a cocoon which it spins around itself. After a lapse of about two months the adult fly emerges, to live for only a few days, just long enough for mating to take place and the eggs to be laid in sand. These hatch in a few weeks to produce the next generation of larvae.

The order Mecoptera is a small group of insects commonly known as scorpion-flies because they tend to carry the slender hind end of the body bent forward over the back in the manner of a scorpion. The larva is eruciform, and has a pair of powerful mandibles designed for chewing animal food. It is not known whether scorpion-flies actually kill their prey or merely feed upon dead insects that they happen to come across.

The caddis-flies, forming the order Trichoptera, are quite a flourishing group. Adult caddis-flies are not unlike moths in general appearance. Their colouring is rather sombre—dull brown, buff or black. Like moths, too, they are mainly nocturnal in their habits, many of them lying up during the day in plants and bushes beside lakes, rivers and streams, or hiding in bark crevices. A dense covering of hairs on wings and body enhance their likeness to moths. Their flight is rather weak, but they are attracted to light, and will often enter houses when the lights are on and the windows left open. Adult life, as in so many other insects, is brief. The mouth parts are much reduced, and there are no mandibles. Many of them do not feed at all, while others can only lap up liquids. Caddis-flies are widely used by fly fishermen, to whom they are known as sedge-flies.

The females lay their eggs either in water or attached to plants growing above the water. Those laid in water may be merely dropped into it by the female, or attached by her to submerged vegetation or stones. Those whose larvae inhabit running water are always securely attached to prevent them being swept away by the current.

The vast majority of caddis larvae are completely aquatic, this stage in the life history lasting for about a year. The larvae themselves are very similar in appearance and general structure to the caterpillars of butterflies and moths. Many of them build the familiar caddis cases in which they live. These cases consist of a lining of silk to which various materials are attached. Each kind of caddis larva usually has its own characteristic type of case, which is always covered with similar material. This may be almost anything occurring in the water—pieces of leaves, stalks, sticks, sand, gravel and even the shells of young water snails, which are sometimes incorporated in the caddis cases while their unfortunate inhabitants are still inside them. The silk is produced by special glands which open into a spinneret on the floor of the mouth.

The cases are tubular and open at each end, the front opening being the larger so that the head and thorax can be protruded to allow the larva to move along the bottom. The covering of the head and thorax is well developed, as are the three pairs of thoracic walking legs. If danger threatens the head and thorax can be rapidly withdrawn into the safety of the case. The larva is attached to its case by means of a pair of claspers situated at the hind end of the abdomen. So efficiently do they perform their function that the larva cannot be forcibly removed from its case without doing serious damage to its body.

Despite the physical protection and camouflage that their cases provide, caddis larvae nevertheless form an important item in the diet of trout. One species, known to fishermen as the brown sedge, seems to have solved the problem of fish predators. It constructs a characteristic case of sand grains with which are also incorporated a number of quite long sticks which project out at various angles. These apparently make too uncomfortable a mouthful for the fish, so that these larvae are able to wander about unmolested in broad daylight.

Certain kinds of caddis larvae that live in running water do not construct cases. Instead they fix themselves to the undersides of stones or on plants within a net made of silk. This net is always constructed so that its open end faces upstream, so as to trap small organisms being carried downstream in the water. These organisms of course provide the larvae with their food. The case builders are mainly plant feeders.

The larval stages take about a year to complete, and then they prepare to pupate. In many species the first stage is the fixing of the case to a submerged stone, so that it cannot be swept away when the active larva is replaced by a passive pupa. At the same time both openings are covered over with a network of silk to prevent entry of intruders while still allowing access to the respiratory current. Pupation in most species is completed in a few weeks, but in some the pupa remains dormant within its case right through the winter, the adult not emerging until the following spring. Even those species whose larvae are not enclosed in a case do form one of stones just before pupation and attach it firmly beneath a large stone.

Caddis pupae are provided with a pair of powerful mandibles with which they are able to bite their way out of the case when the time comes for them to emerge. At the conclusion of pupal life the pupae either climb out of the water or swim up to the surface. The adult insects then burst out and fly away. In many insects the newly-emerged adult must rest while its wings dry and harden, but many of the caddis-flies are able to fly the moment they emerge. This is particularly so with those species that live in running water, the fly bursting out of the pupal case and taking to the wing the moment the swimming pupa breaks the water surface.

Butterflies and moths are among the best known of all the insects. The order Lepidoptera which they constitute is among the four most successful of all insect orders. The great majority of them are moths. The feature that distinguishes the Lepidoptera from all other insects is that their bodies and wings are covered with flat scales which come off on the fingers like fine powder if they are handled. Each scale has a short stalk which fits into a socket, and they are arranged in neat rows that overlap like the tiles on a roof. So far as is known these scales do not help in flight in any way, nor are they protective. Apparently their main purpose is to provide their owners with their colour, although some of them can be modified as scent glands.

Both pairs of wings are well developed, and there is an important difference in arrangement between the wings of butterflies and those of moths. Butterfly wings are held together in flight because the front pair considerably overlap the hind pair, but in the moths there is a special locking device to hold the wings together. A hook-like catch near the base of the fore wing receives a stout bristle projecting forward from the base of the hind wing.

Butterflies and moths differ, too, in their antennae. Butterfly antennae consist of a long smooth shaft ending in a knob, whereas moth antennae are more variable, many of them being feathery, and none of them ending in a knob.

Another lepidopteran feature is that in many species the

two sexes are quite different in appearance, often so much so that no one would suspect them of belonging to the same species. In some moths these sex differences extend beyond colour and size, the males being winged while the females are without wings and therefore unable to fly. It would seem, though, that loss of flight by the females is not a real handicap to these species, for many of them are widespread and common.

In most species the females lay their eggs singly, each one usually being deposited on a different plant, presumably to ensure that the caterpillars have ample supplies of food after they are hatched. A minority, however, notably the cabbage white and the tortoise-shells, lay their eggs in batches. Most caterpillars are very specific in their diet, many feeding on only one kind of plant, while others confine their attentions to a few related species. In consequence the females lay their eggs on these plants, thus ensuring that the young caterpillars when they hatch will have supplies of suitable food immediately available without having to travel in search of it.

Butterfly and moth eggs vary very considerably in shape. They all however have a whitish or transparent shell made of chitin, at the top of which there is a minute hole known as the micropyle. It is through the micropyle that the spermatozoon enters to fertilise the egg as it travels down the female oviduct. It also allows supplies of air to reach the developing embryo.

Within a week or two of laying the young caterpillar is normally ready to hatch. This it does by biting its way out with its already well-developed mandibles. As soon as it has emerged it proceeds to eat the egg shell. This appears to be necessary for some species, which die if it is removed before they can eat it. As soon as it emerges it begins to eat, and as it grows it must moult periodically. Lepidopteran larvae moult between five and nine times before they are ready to pupate, the number of moults usually being constant for any given species.

The head of the caterpillar has a hard cuticle to provide anchorage for the strong muscles that work its powerful jaws.

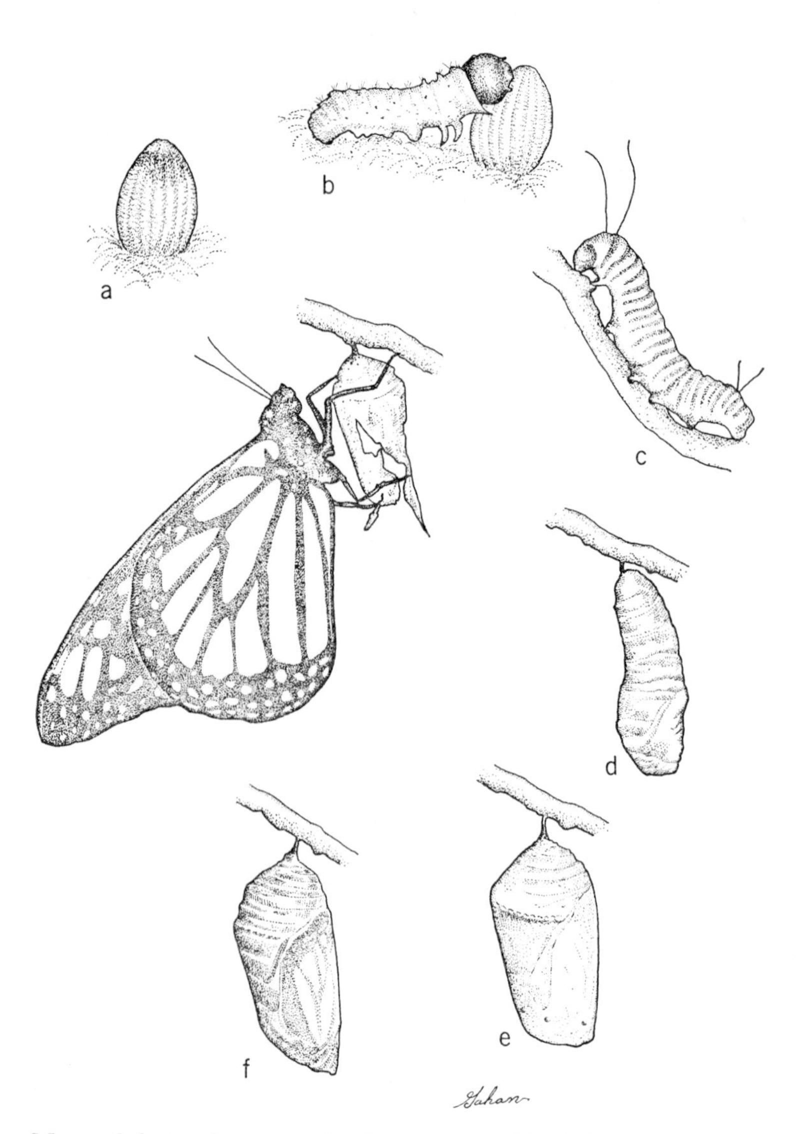

Monarch butterfly: *a* egg glued to milkweed leaf; *b* newly hatched caterpillar, about to devour its eggshell; *c* grown caterpillar; *d* pupa; *e* pupa within case; *f* two weeks later, wings showing through transparent case; *g* new-born monarch, blood pumped into its wings, ready to fly

The larval thorax consists of three segments, each bearing a pair of legs ending in a claw, and these are followed by ten abdominal segments. In the majority of species the third, fourth, fifth, sixth and last of these each bears a pair of pro-legs or false legs, which are purely larval structures enabling the caterpillar to move about. They are not represented in the adult. Pro-legs are soft and fleshy, and each ends in a pad surrounded by a ring of hooked hairs which enable it to grip the surfaces on which the caterpillar walks. In one group of moths, the geometer moths, the caterpillars are unusual in that they possess pro-legs only on the sixth and the last abdominal segments. In consequence they move by a looping movement—these are the familiar looping caterpillars.

The vast majority of caterpillars live on leaves but a few live underground, where they feed on roots. The larvae of all the swift moths have this habit, and consequently do considerable damage to cultivated plants, especially those of the herbaceous border.

The most unusual lepidopteran larvae are those of a family of moths known as the Pyraustidae. They are completely aquatic, and are not easy to find because they construct cases similar to those made by caddis larvae, merely extending their heads beyond the open front end when they want to feed. These cases are made from pieces of their food plants which are bitten off and fastened together. The Pyraustidae are commonly known as china moths, because the delicate patterns on their wings are reminiscent of the marks found on the base of good porcelain.

The caterpillars of certain kinds of moths live in communal webs constructed in trees and hedgerows. Each web may well contain the progeny of more than one female. By the time they have reached their last larval instar they have begun to disperse, so that they pupate as solitary individuals.

At the end of larval life the lepidopteran caterpillar moults for the last time, and the new skin hardens to a rigid coat within which pupal development takes place. This involves a drastic rearrangement of the larval organs in which they are converted into the organs of the adult. During this time the pupa does not feed, and all its energy requirements are provided by the food materials laid down during larval life. Eventually, within the pupal case, the typical caterpillar has been reconstructed into a typical adult or imago.

In most moths the last larval instar covers itself in a cocoon of silk or similar material before pupating, but only a few butterflies construct such a cocoon. The best known of these cocoons are of course those produced by the silk moths, whose larvae are the familiar 'silkworms'. The domesticated silkworm is the larva of the moth *Bombyx mori*. Each cocoon is said to produce a strand of silk a thousand feet long. Despite this something like twenty-five thousand cocoons are needed to provide one pound of pure silk!

A warm sunny day even before spring has officially begun may well bring out the first butterflies. These early butterflies, however, will represent only a few species. The others will make their first appearances at various times, some not until well into the summer.

The explanation for these progressive appearances lies in the varying methods of survival through the winter adopted by different species. As autumn approaches the adults of some species go into hibernation, and these are the species that appear first in the spring. Many of these early specimens look rather frayed after their long winter in hiding. Some individuals manage to hide away indoors, and it is these that make sudden appearances fluttering in the windows.

A few butterflies pass the winter as hibernating pupae, and a few more as hibernating eggs, but the majority hibernate as larvae. In some species the caterpillars have grown almost to full size before they hibernate, and will pupate shortly after coming out of hibernation the following spring, while others have varying proportions of their larval life still to complete. In extreme cases the tiny caterpillars hibernate as soon as they have hatched. Before they do so, however, they eat their egg cases, but take no other food. These species, and those that over-winter as eggs, are the last to appear the following year.

Whenever they emerge adult butterflies and moths are much concerned with mating and egg-laying. Males may find their mates by sight or by scent. All female lepidopterans that are ready to breed give off a scent that is attractive to males. The scent produced by female butterflies is believed to be effective over distances of up to a hundred yards, but since butterflies are diurnal the coming together of the pairs probably depends more upon sight than scent. But with moths, which are mostly nocturnal, this attraction by scent probably plays a much more important part. In some moths the male can be attracted to a female from as much as three miles away. In these species the feathery male antennae are very well developed, and act as highly efficient sense organs in the detection of female scent. The female produces her attractive scent from organs situated near the tip of her abdomen. As many as fifty males have been known to be attracted to a single virgin female. It is interesting to note that only unmated virgin females are capable of exerting this attraction for the males. Once a female has been mated males take no further interest in her. Even an empty box that has contained a virgin female can exert the same attraction.

That it is their antennae that enable them to detect the virgin females has been proved by coating the antennae of males with shellac or some other material. Under these circumstances they show no interest in virgin females even when they are within sight.

Once a pair have come together through the scent produced by the female the stimulus to mate is provided by scent produced by the male. This scent is produced by glands situated in the wing membranes and is disseminated by the scales that are immediately above them. These scent-distributing scales are known as androconia.

It is rather interesting that, although we cannot detect the female scent of any species, the scents emitted by many male butterflies and moths are easily perceptible, and they are invariably attractive to us as well as to female butterflies and moths. There is considerable variety in these scents, which are variously reminiscent of chocolate, heliotrope flowers, meadow-sweet, sweet peas and old cigar boxes. These scents are detectable only in a certain proportion of the males of any species, presumably because they are produced only while they are in active breeding condition.

Flies, constituting the order Diptera, are found in almost every part of the world in a great variety of types. The feature that distinguishes them from all other insects is that they have only a single pair of wings, representing the front pair of the two pairs present in most other insects. The hind pair is represented by two small knobs borne at the ends of short slender stalks. These halteres, as they are called, are sense organs which play an important part in enabling the fly to maintain its balance when it is flying. If they are removed normal flight becomes impossible, and the fly falls to the ground.

Dipteran larvae are apodous, none of them developing thoracic legs, though many of them, especially those that live in water, do develop abdominal pro-legs. The most advanced type of larva is the maggot, which has a pointed head and a blunt hind end, and is terrestrial in its habits. When the last larval instar pupates the larval skin is retained as an extra outer coat which becomes hard, dark and barrel-shaped. It is in fact a type of cocoon known as a puparium. When metamorphosis is complete and the young adult fly is ready to break out it has to rupture this puparium as well as the normal pupal skin. For this purpose it is provided with a special frontal sac at the front of the head. This is filled with blood under pressure, which causes the top of the puparium to rupture, allowing the adult fly to emerge. The frontal sac is then withdrawn, and is of no further use to the fly during its lifetime.

One of the best known and most widely distributed of all flies is the common house fly, *Musca domestica,* which is absent only from very cold areas. Its habits, too, make it one of man's worst enemies among insects. Dung of all animals, including humans, has a great attraction for it, and while feeding on it its feet and mouth parts become contaminated not only with the dung but with any disease organisms it may contain. From this unwholesome feast it may well fly straight into a house or a shop and settle on any uncovered food, at best depositing filth on it, and at worst infecting it with any germs it may have picked up. In various parts of the world cholera, typhoid, dysentery and summer diarrhoea are disseminated by the house fly. Its eggs will be laid in any accumulation of dung, but it has a preference for horse dung. The reproduction rate is high. Even in temperate climates, where summer temperatures are not high, within three weeks of hatching adult flies may themselves be laying eggs to produce the next generation.

The most abundant relatives of the house fly are the various kinds of blowflies, the bluebottles, various species of *Calliphora*, and the greenbottles, species of *Lucilia*, to which must be added the fleshfly, *Sarcophaga carnaria*. Bluebottles and greenbottles lay their eggs on meat, including of course the corpses of any animals they come across, and their maggots are the gentles known to every fisherman. A meat safe or a gauze cover is sufficient to protect the weekend joint from becoming infected by blowfly eggs, but it is not proof against the larvae of the fleshfly.

Instead of laying eggs that will eventually hatch into maggots, the female fleshfly retains her fertilised eggs within her oviduct, so that what she deposits is not eggs but first instar larvae. Normally these are released on dead animal bodies wherever she can find them, but if she finds meat protected beneath a gauze cover all she has to do is to settle on the top of the cover and drop her larvae through the mesh on to the meat.

When the last larval instar becomes full grown it proceeds to burrow into the earth six inches or more below the surface before pupating. The subsequent release of the adult fly is quite a complicated process. First it emerges from the puparium by inflating its frontal sac as already described for the house fly. But it now has to negotiate the several inches of soil above it before it can expand its wings. This it does by continuously expanding and contracting its frontal sac. As it expands it compresses the soil immediately above the fly, thus enabling it to move a short distance upwards. The sac is then thrust into the next layer above and expanded. Thus the fly slowly moves upwards until eventually the last expansion of the frontal sac breaks the surface to free the fly. Its first task is to brush off all the soil particles from its body, wings and frontal sac, using its legs, after which the frontal sac is withdrawn and the wings are expanded and hardened.

Blowfly maggots are bred on a commercial scale to supply the needs of fishermen for bait. The flies themselves, as well as house flies, are bred in large quantities in many zoos for use as food for lizards and other small reptiles. When the maggots pupate they are put into cold storage, from which they can be removed to emerge as flies as they are required.

A number of different kinds of flies attack cattle and other domestic animals. The horse bot fly lays its eggs on the skin of horses, always choosing places that are within reach of the horse's mouth. When they hatch they cause considerable irritation. The horse responds by licking the affected part, and in doing so swallows the newly-hatched larvae. These attach themselves to the lining of the gut, where they remain until they are fully developed, which takes between nine and ten months. Finally the last larval instars release their hold and pass out with the horse's dung. They immediately bury themselves in the ground and pupate, emerging as adults after about six weeks.

The related sheep bot fly is viviparous, the female depositing first instar larvae in the nostrils of sheep, from where they travel to the safety of the frontal sinuses. Here they remain for about nine months attached by a pair of mouth hooks to the mucous membrane and feeding on the mucous. The larvae are deposited in the nostrils during the height of the summer, so the final instar larvae are ready to emerge from their safe hiding places the following spring. Accordingly they release their hold and are sneezed out by the sheep. Soon after their emergence they pupate beneath stones or tufts of grass. When the adult finally appears it lives for only a short time, during which it does not feed. Its mouth parts in fact are only rudimentary and not functional.

The ox warble fly is an equally unpleasant creature. The female attaches its dozen or so eggs to hairs on its victim's legs and flanks. Each egg case has a clamp-like extension which serves to anchor it to the hair. When the maggots hatch they bore their way beneath the skin, and over the next two months or so work their way through the host's body until they finally arrive beneath the skin of the back. Here each larva forms a small perforation of the skin which serves as a breathing pore, beneath which a large tumour-like swelling is produced. This is known as a warble, and it contains quantities of pus produced by the host in response to the irritation produced by the larva, and it is on this pus that the larva feeds. When it is finally ready to pupate the last larval instar forces its way out of the breathing pore and falls to the ground, in which it pupates. About a month later the adult warble fly emerges

One of the most interesting of the flies that attack mammals is the sheep ked, sometimes known erroneously as the sheep tick, for it is a true insect and not a tick. It lies as a parasite in the wool of sheep, sucking their blood. Both sexes are without wings, and the female produces only one egg at a time. This she retains within her oviduct while it undergoes all its larval instars. Only when the last instar is ready to pupate does it emerge, whereupon it pupates immediately, transforming itself in a few weeks into an adult ked.

Among the most common insects to be found wherever flowers are blooming in the summer, especially members of the Compositae or daisy family, are the hover flies, comprising the family Syrphidae. Although as adults they feed on nectar, the larval stages of the majority of them live as predators, aphids being their chief source of food. The adult females of these kinds lay their eggs on plants where flourishing colonies of aphids are already established. From these eggs emerge flattened, slug-like larvae which are pale in colour blotched with green or brown. They vary considerably in size, those of some of the larger species exceeding half an inch in length. Like slugs, too, they produce a mucous-like lubricant which enables them to glide over the surface of leaves in search of their prey. When an aphid is seized the whole of its soft interior is sucked out, leaving only its shrivelled skin. The value of these hover flies in keeping aphids in check can be gauged by the fact that observation on captive specimens has shown that a single larva may be capable of devouring as many as nine hundred aphids before it pupates.

A minority of hover flies, however, produce larvae that are quite different from these both in appearance and habits. The females of these species lay their eggs in stagnant water in which there are plenty of leaves and other decaying organic material. From them hatch some of the most unusual of all aquatic creatures, the rat-tailed maggots. These live half buried in the mud beneath shallow water, feeding upon such organic materials as it may contain. They are between half an inch and one inch in length, but the hind end is extended as a long breathing tube made up of three segments which can be telescoped or extended as required. Fully extended, this breathing siphon may be as much as six inches in length. It passes straight up to the surface of the water, where it can take in the air necessary to keep the submerged larva alive.

At its tip this breathing siphon is provided with a circle of eight fringed hairs, and when it is at the surface of the water these hairs spread out to form a rosette which lies on the water. Their function is to prevent water entering the siphon, and they are able to do this because they are incapable of being wetted. As the rat-tailed maggots move about on the bottom, so their depth below the surface varies, and the lengths of their siphons are adjusted accordingly so as always just to reach the surface but not to project beyond it. If a maggot goes so deep that the tip of the siphon remains below the surface it soon moves into shallower water so that the siphon can reach the surface again.

When these maggots pupate they retain their general larval shape, including the siphon, but this is now functionless.

Breathing is taken over by a pair of short breathing horns which project upwards from near the front end of the body. The pupae float just beneath the water surface, enabling these horns to project above it.

Another important group of flies that have aquatic larvae are the mosquitoes. There are two types of mosquitoes, the anopheline and the culicine, and there are significant differences between the larvae of the two types. Anopheline larvae lie horizontally just beneath the surface of the water, respiratory air being taken in through a pair of spiracles situated on the eighth abdominal segment. In culicine larvae, however, only the tail end comes near the water surface, air being taken in through long breathing tubes which project upwards from this end of the body, the rest of the body hanging downwards in the water. The anopheline spiracles and the culicine breathing tubes are both covered with an oily secretion that is water repellent, so preventing water from entering the tubes. This has an extremely important bearing on methods of destroying mosquitoes.

The differences in their orientation at the surface play an important part in their methods of feeding. Both kinds of larvae are provided with moustache-like feeding brushes on either side of the mouth. These vibrate rapidly as they gather minute food particles and pass them towards the mouth. Culicine larvae gather their food from a little below the surface. An anopheline larva, however, turns its head through about one hundred and eighty degrees when it is feeding so as to be able to sweep up food particles adhering to the surface film. If disturbed both kinds of larvae are able to relinquish their adherence to the surface and swim quite rapidly through the water by means of wriggling movements of the body.

Insect pupae in general are unable to move, but mosquito pupae are unusual in that they are very active. In general appearance they are somewhat like tadpoles, having a large head and thorax combined with an active abdomen. From the upper surface of the thorax a pair of breathing trumpets projects into the air at the surface of the water, enabling the pupae to breathe. This pupal stage is very short, the adult mosquito usually emerging within a week. The actual emergence, too, is quite rapid, occupying only about five minutes.

Even in their egg-laying the anopheline and culicine mosquitoes show important differences. Culicine mosquitoes lay their eggs in great numbers, sometimes as many as three hundred at a time, and these adhere together to form a raft of eggs which floats at the surface of the water. Anopheline eggs are laid singly, but these too remain at the surface of the water because each is provided with a pair of air-filled floats.

Whereas in most insect orders there is a considerable variety of types, the fleas, which form the relatively small order Siphonaptera, are a remarkably homogeneous group. They are all similar, both in their structure and in their habits. The mouth parts of the adults are modified for piercing and sucking. All fleas are parasites and feed on the blood of mammals and birds, and they are completely devoid of wings. Their eggs are laid in the fur or feathers of their hosts, but they do not undergo development there. They, or the larvae that hatch from them, fall off when the hosts are in their dwelling places, and here the larvae feed on any organic debris found in the dwellings. To enable them to chew up this food their mouth parts are adapted for biting. Because the breeding site is as important to them as the adult host, fleas are generally parasitic only upon mammals and birds that have some kind of nest or lair to which they return from time to time. They are not found, except as occasional strays, on nomadic animals like cattle and horses which do not have lairs.

CHAPTER FIFTEEN

INSECTS : PART THREE

In this chapter we come to what are perhaps the two most successful orders in an extremely successful group—the order Hymenoptera, and the order Coleoptera, which comprises the beetles and weevils and contains many more species than any other insect order.

The name Hymenoptera refers to the well developed membranous wings that the members of the order possess. There are two suborders, the Symphyta and the Apocrita. The former are the primitive members of the order, their principal members being the saw-flies and the wood-wasps. Their larvae are almost exclusively plant feeders, and are similar to the caterpillars of the Lepidoptera, having thoracic legs. Many of them also have abdominal pro-legs. The suborder Apocrita contains a great variety of types and is divided into two sections, the Parasitica, containing the gall-wasps, ichneumon-flies and chalcid-wasps, and so named because the vast majority of their larvae are parasitic; and the Aculeata, or stinging Hymenoptera. These are the wasps, bees and ants, which have evolved a highly organised social system, and whose ovipositors have become modified as stings.

Saw-flies are so named because the ovipositor of the female is serrated. With it she is able to make slits in plant stems, and in these she deposits her eggs, so that they are protected while they develop. After hatching the larvae emerge from their hiding places and feed upon the foliage of the plant on which they find themselves. Each species generally has its own food plant on which the female deposits her eggs.

Parthenogenesis is a common feature of many of the Hymenoptera; and at least in some saw-fly species unfertilised eggs produce males and fertilised eggs develop into females. Presumably under normal conditions the females lay an assortment of fertilised and unfertilised eggs to maintain the balance of the sexes. As the last larval instar pupates it forms a cocoon, which may be attached to the food plant but is more usually buried in the ground.

In contrast to the saw-flies, which are small and insignificant insects, the wood-wasps are quite large. The giant wood-wasp, *Sirex gigas*, one of the best known species, has a body about one and a half inches long, and the female has a stout ovipositor about half an inch in length. With this she is able to bore a hole in the trunks of conifers, and at the end of this hole she deposits an egg. The larva that hatches remains in the trunk for two or three years, and during this time it works its way through the wood, frequently penetrating right into the heart wood. Large numbers of wood-wasp burrows seriously reduce the value of the timber when the tree is felled.

However far into the trunk the larva may burrow, before the last instar is ready to pupate it comes near to the surface of the trunk, and pupates so that when the adult emerges its head is facing outwards, and it only has to bore its way ahead to be released. How the larvae manages to orientate itself in this way is unknown.

There is some mystery, too, about the way the larva nourishes itself during its long imprisonment. It is thought that it does not in fact eat wood but the products of the action of moulds on the wood. When the females lay their eggs they also deposit from a pair of special glands at the bases of their ovipositors mould spores, which germinate to form moulds which spread along the tunnels excavated by the larvae. It seems likely that these moulds digest the wood and that the larvae feed on their products of digestion.

The remaining members of the Hymenoptera are distinguished from the Symphyta, and indeed from all other insects, in their possession of the typical 'wasp waist'. The first abdominal segment is fused with the thorax, and the second is much constricted and often elongated to form the 'waist'. The larvae are legless grubs, but in contrast to the maggots of flies they have a well defined head.

Like the saw-flies and wood-wasps, the gall-wasps also deposit their eggs in plant tissues. But whereas plants seem to tolerate the presence of the larvae of the first two, they react to gall-wasp larvae by producing galls, as if in an attempt to seal off irritating intruders. It is a curious fact that oak trees are much more prone to gall production than any other plant. Why this should be is a complete mystery.

The two best known oak galls are the soft oak-apple and the hard spherical marble gall. Oak-apples develop as terminal swellings of twigs as the leaves begin to grow in the spring. By the beginning of May they are already obvious, and by the end of the month they are well developed. Each oak-apple consists of a number of separate chambers, each occupied by a grub of the gall-wasp, *Biorhiza pallida*. After pupation the adult wasps emerge in July by boring neat round holes up to the surface. The males are provided with fully functional wings, but none of the females can fly, some of them being entirely without wings while the others have tiny functionless wings. After mating the females crawl down the trunk and into the ground. Here they lay their eggs on the tiny lateral roots of the tree, inserting them in slits which they make with their ovipositors, each slit receiving a single egg. As the eggs develop they stimulate the formation of small galls on the roots.

By the end of the summer the life history has been completed and the adults emerge. These are all wingless females, no males appearing at this time. They proceed to crawl out of the ground, up the trunk and out to the terminal buds of twigs, where they lay unfertilised eggs. These remain dormant until the following spring, when they begin to develop and cause the buds to swell, producing the oak-apples.

The hard marble galls are caused by the reaction of the oak

tree to the developing larvae of the gall-wasp *Cynips kollari.* Each gall never contains more than one grub. In this species parthenogenesis seems to have achieved its ultimate success, for no male of this species has even been found. Further investigation will reveal whether this particular species has in fact succeeded in finally eliminating the male.

The ichneumon-flies are all parasites in their larval stages. The females lay their eggs either on or in the tissues of other insect larvae. The larvae that hatch from these eggs feed on the host larvae, and pupate within their shells.

Chalcid-wasps are also parasitic in their larval stages. They are much smaller than ichneumon-flies and lay their eggs on the eggs rather than the larvae of their victims. And the subsequent development of these eggs in some species reveals a remarkable phenomenon. Many chalcid-wasps lay their eggs in the eggs of various species of moths. As the moth eggs hatch, so also do the chalcid-wasp eggs. But instead of developing at the expense of the moth eggs to form single larvae, they divide to form a large mass of embryos which may number hundreds or even thousands. In an extreme case no fewer than three thousand adult chalcid-wasps have been recorded as emerging from a single larva of the silver Y moth, *Plusia gamma.* This phenomenon, whereby a single egg can give rise to large numbers of offspring, is known as polyembryony.

In the Aculeata or stinging Hymenoptera it is the ovipositor that is modified as the sting, which is therefore present only in the females of the species. It can of course still function as an ovipositor, the eggs passing out through an opening at its base. Although the bees, wasps and ants that make up the group are regarded as social insects, it is in fact only in the ants that an elaborate form of social life is found in all species. The majority of bees and wasps are solitary, only the minority showing a degree of social organisation comparable with that found in the ants.

Among the bees the social ones are the hive or honey bees and the bumble bees. Although they form only a tiny minority of the total bee species, they are much better known than the solitary bees because there are many more of each kind. One of the most characteristic sounds of spring is the rich low hum of the large bumble bees that appear as soon as the early shrubs come into bloom.

These spring bees are queens or fertile females which have spent the winter in hibernation as the sole link between the populations of one year and the next. Late the previous autumn all the workers and males died off while at the same time the young queens, recently emerged and mated, sought out suitable burrows in the ground or other safe places in which they could sleep through the winter protected from the worst of the weather.

The task facing each of these newly-awakened queens is an urgent one. She must first find a suitable site for a nest, and this usually means a hole previously occupied by field mice or voles, but now abandoned. Some species instead prefer to nest in the base of a tussock of grass, these species being known as carder bees because they make their nests by plaiting together dried grass and other plant material. Having chosen her nesting site, the queen bumble bee proceeds to make one large wax cell in which she then lays a dozen or so eggs. Nearby she also constructs a honey-pot, and it is while she is filling this that we see her so active around the spring flowers. A bumble bee honey-pot may hold as much as a thimbleful of honey.

The mouth parts of bees are adapted for licking up fluids, and in the social bees they are greatly elongated so as to be able to reach right to the bottom of flowers, from which they can gather nectar. Nectar is a solution of sugars. This is swallowed by the bee, and in its crop it is acted upon by enzymes to convert it to the simple sugars dextrose and laevulose. Finally the fluid, which is now honey, is regurgitated into specially constructed cells. These simple sugars have the great advantage for feeding to the developing larvae that they do not need digesting.

From the original eggs the first batch of workers appears, and they now take over the duties of looking after the subsequent broods, leaving the queen to concentrate on her main job of egg-laying. Thus it is that by late spring all the very large bumble bees have virtually disappeared, for the workers that are now actively collecting nectar are much smaller creatures.

Although bumble bee nests never attain the size of wasp nests, they may nevertheless hold up to four hundred individuals at the height of the season. But not until late in the summer will any males appear. Their sole function is to mate with and fertilise the young queens which appear at the same time. These are the queens that hibernate through the winter. The spermatozoa obtained at mating are stored in a spermatheca, which is a sac leading off the oviduct. Here they will remain alive and will be available for fertilising all the eggs laid during the following year. No queen bee ever mates for a second time.

The social bees, both honey bees and bumble bees, have an interesting method of regulating the production of queens, workers and males. As an egg passes down her oviduct the queen seems to be able to decide whether to release a spermatozoon from the spermatheca to fertilise it or not. If it is unfertilised the resulting bee will be a male, while a fertilised egg generally produces a worker, which is really a sterile female. Queens, which are of course fully fertile females, also result from fertilised eggs. In the honey bee, queen instead of worker production seems to be achieved by feeding the queen grubs larger quantities of richer food than is given to potential worker grubs (this food, rich in protein, is known as royal jelly). To what extent more or richer food induces the production of bumble bee queens is not yet known.

The characteristic cells in which the social bees not only lay their eggs but also store honey and pollen are made of wax, which is produced from glands situated between the abdominal segments. In the honey bee these cells are constructed with mathematical precision, each being a perfect hexagonal prism, whereas those of the bumble bee are roughly spherical in shape. Each bumble bee brood cell receives a variable number of eggs, usually between eight and fourteen, but no honey bee cell contains more than one egg. Before she lays her eggs the queen bumble bee furnishes the brood cell with considerable quantities of pollen, the result of many foraging expeditions. Having smoothed it down so that it forms a floor to the cell she then lays her eggs on it and covers the cell with a wax roof.

As soon as the larvae hatch they begin feeding on their pollen store. But they also need honey, and this the queen provides by periodically biting a hole in the wall of the cell and regurgitating honey from her honey-stomach into the cell before sealing up the hole. When fully grown, the last larval instars spin silk cocoons around themselves and pupate.

In the honey bee the normal brood cells come in two sizes, larger drone brood cells measuring about one-quarter of an inch across and smaller worker broods cells about one-fifth of an inch across. At certain times very small numbers of much larger queen brood cells are constructed. Although the drone and worker brood cells may be mixed up in the same comb, the queen bee seems to be able to lay unfertilised eggs in the drone cells and fertilised eggs in the worker cells with uncanny accuracy.

These honey bee brood cells are not capped like those of the bumble bee, and as soon as the larvae hatch the task of feeding them is taken on by the workers. Only when the last larval instar is about to pupate is a wax lid put over each cell.

In addition to their lack of social organisation, solitary bees also differ from the colonial kinds in their inability to make wax. Instead they make their brood cells from a variety of materials. Mason bees, for example, construct cells made from a mixture of earth, sand, tiny pebbles and even minute particles of wood, glued together with a secretion produced by their salivary glands. Leaf-cutting bees construct their brood cells from neatly cut out pieces of leaves and petals. They are particularly fond of rose leaves, and if you have a rose garden you may well be able to see the results of their activities. Most solitary bees excavate burrows in the ground for their nests, though almost any suitable crevice may be used. Many of the solitary bees are parasites which live in the nests of other solitary bees.

As with bees, the great majority of wasp species are solitary in their habits, social life being the exception rather than the rule, but it is with the minority of social species that most people are more familiar. The most significant difference between wasps and bees is that, whereas the latter feed exclusively on plant products, except for a few parasitic species, all wasp larvae are carnivorous, the food provided for them consisting of other insects and their larvae and other small invertebrate animals. Adult wasps, however, are unable to swallow solid food, so they exist in the main on plant juices. Some of them also lick up animal juices, usually from dead animals.

No wasp is able to produce wax, but they are very skilled at making hexagonal cells similar to those constructed by the social bees out of 'wasp paper'. This really is similar to manufactured paper. To prepare it the wasp rasps off minute particles of wood from trees, posts and fences with its powerful mandibles, and then works them up into a kind of papier mâché by mixing it thoroughly with saliva.

In their social organisation wasps are more like bumble bees than honey bees. Their colonies are annual affairs, all the workers and males dying off in late autumn, leaving the young fertilised queens to hibernate through the winter. These are the large wasps which are so familiar when they emerge from hibernation in early spring. The queen wasp's first task is to search for a suitable hole in the ground in which she can begin to build her nest. A deserted mouse hole makes an ideal site. She must now work hard gathering wood for paper-making. The first of this paper is used to form a disc-like roof to the cavity, and from this roof a paper stalk is constructed hanging downwards into the cavity. A horizontal layer of hexagonal cells is now constructed attached to this stalk, each cell being closed above but left open at the bottom. As this first comb extends laterally, so a paper canopy is constructed to cover it by extending the initial roof disc.

As the cells are formed an egg is laid in each one, being

glued to the inside of the cell to prevent it from falling out. All of these eggs are fertilised and so eventually produce workers, sex determination in wasps being governed in exactly the same way as it is in the social bees. As soon as these first workers emerge they take over the collection of food and paper and the construction of more cells, leaving the queen to devote herself entirely to her main function of egg laying.

As the nest grows, so soil has to be excavated and dumped outside away from the entrance to the nest to make room for the ever increasing numbers of brood cells. When the first comb has been extended sufficiently a second comb is begun beneath it, attached to the first, as the first was to the roof disc, by a pendant stalk. When it is fully formed this second comb is attached to the first by a whole series of such stalks. In a large nest there may be finally as many as eight or even more of these horizontal combs suspended one below the other.

Until late summer only worker wasps are produced, but now the queen lays some unfertilised eggs, which will give rise to males, and the workers construct some larger brood cells. In these she lays normal fertilised eggs, identical with those that have been producing workers, but the grubs that hatch from them are fed larger quantities of food richer in protein. After pupation they emerge as the new generation of queen wasps, and are fertilised by the males hatching at the same time. Soon after this the males and all the remaining workers die off with the approach of winter, leaving the queens to go into hibernation until the following spring.

A successful nest may contain well over ten thousand cells, but not all the wasps produced from them will be present in the nest at the same time. Those that appear in the earlier part of the season will have completed their work and died before the later ones appear. Even so, a nest in September will often have well over three thousand active workers.

All wasp grubs are fed on pieces of insects and other small animals which are caught and cut up by the active workers. Thus even a single wasp nest will account for an enormous number of undesirable insects during the course of the summer. The wasp, then, despite its sometimes irritating presence, is certainly a beneficial insect; and killing queen wasps in the spring, and later destroying wasp nests (except when they are very near to dwellings) is a very short-sighted policy.

Although there are many different kinds of solitary wasps they all share one characteristic. Their eggs are laid each in a separate chamber or cell, and are provided with the dead or paralysed body of an insect or other small invertebrate. When the egg hatches it has sufficient food to carry it right through to pupation. In some species the female is so skilful that she is able to penetrate the main nerve ganglion of the victim with her sting. This paralyses the prey but does not kill it, so that it remains alive but helpless awaiting the hatching of the wasp egg whose larder it is destined to become. In other species the prey is killed, but in this case the venom injected by the sting acts as an antiseptic, preventing the body from decaying before the larva hatches and is ready to feed on it.

There are two main types of solitary wasps, the potter wasps and the digger wasps. Potter wasps construct vase-shaped cells made of earth and attach them to the stems of low-growing shrubs such as heather. Most of them stock their brood cells with larvae, some species choosing the caterpillars of butterflies and moths, while others prefer the larvae of beetles.

The more numerous digger wasps excavate brood cells in the ground. There is a great variety in the choice of victim, but any particular species usually confines its attentions to one type, and the majority choose adults rather than larvae. Among the more common victims are flies, weevils, bees, aphids, spiders, caterpillars and the nymphs of frog-hoppers.

The ants are the only group of insects in which social life is universal. There are no solitary ants, and indeed the ants have achieved a higher degree of social organisation than any other insects. Their colonies, like those of the honey bees, are perennial. Individuals, too, live for an unusual length of time for insects. Worker ants are known to be able to live for as long as seven years, and queen ants for up to fifteen years; and these extended life spans probably explain the high degree of social organisation found in almost any ant colony. In contrast to almost every other kind of insect, ants are essentially creatures of the ground, only bearing wings for a short time during the mating season.

As with social wasps and bees there are three fundamental types or castes of ants. Workers are produced from fertilised eggs whose subsequent larvae are provided with a normal diet. Queens are produced from similar eggs, but the larvae that hatch from them are fed on a richer and more plentiful diet. Males result from the development of unfertilised eggs. The occasional eggs that are laid by the normally infertile workers are not fertilised, and therefore augment the supply of males.

Males and queens are produced only over a very short period at the height of the summer, and when they first emerge they are always furnished with functional wings. In a short time they leave their nests for their nuptial flights. Usually all the nests in an area disgorge these young males and queens at about the same time; thus at this time, which seems to depend upon just the right atmospheric conditions, the air will be thick with them. Mating takes place during flight, after which the fertilised queens return to earth and tear off their wings. From now on their lives will be exclusively subterranean, no matter how many years they may live. Worker ants never have wings.

When a queen returns to earth after her nuptial flight she may do one of three things. She may return to the nest in which she was brought up, or she may join another one where, so far as we know, she will be equally welcome. Alternatively, she may decide to found an entirely new nest of her own, and this is what the majority do. In this case she will excavate a small hole in the ground, closing up the entrance when she is safely inside. Here she remains for some time while the eggs mature in her body. Until they are ready to be laid, which may not be for two or three months, she lies dormant, relying upon her body's food reserves, including the well developed wing muscles for which she has no further use, to keep her alive.

From the first eggs that are laid only workers emerge, and these take over all the duties of nest-building, tending of offspring and foraging for food. The queen from now on devotes all her attention to egg-laying, and is fed by the workers on regurgitated liquid food.

The order Coleoptera, comprising the beetles and weevils, is an immense collection of insects. Already more than a quarter of a million species have been named and described, and several hundred new species are being added to the list every year. This single order in fact contains about one-third of all known insects and about as many species as the whole of the rest of the animal kingdom excluding the insects.

Although they differ very considerably in shape and size, and between them inhabit almost every conceivable habitat, the Coleoptera share one easily recognisable feature. The front pair of wings is always represented by a pair of horny elytra which meet down the middle line and never overlap. They are never used for flight, this being the function of the hind pair of wings, which is usually much longer than the elytra. At rest, however, they are folded transversely as well as lengthwise so that they can be tucked away completely beneath the protective covering of the elytra. The first thoracic segment, the prothorax, is well developed. The mouth parts of all adult Coleoptera and of many of their larvae are strongly developed and adapted for biting and chewing, the mandibles being toothed.

With such a vast assemblage it is possible to choose only some of the better known and more common types in this survey. Among the best known are the ground beetles, extremely predatory insects with long legs enabling them to run swiftly after their prey. Although their wings are well developed, ground beetles seldom fly. They are essentially creatures of the soil, living either actually in the ground or under stones, bark, wood or other objects lying on the ground. The larvae are as active as the adults, with powerful jaws, and equally predatory.

Closely related to the ground beetles are the tiger beetles, which are found in abundance only in the warmer parts of the world. In their habits, too, they are similar to the ground beetles, but their larvae are rather sluggish. Each excavates a vertical burrow which extends about a foot into the ground. It then lies at the mouth with only its head and prothorax protruding, waiting for some unwary insect or other small animal to wander within range. When this happens the larval body jerks into action and the prey is seized in its mandibles. The larva then retires to the bottom of its burrow to consume its meal at leisure. It maintains its watching position at the mouth of the burrow by means of its powerful legs and a specially developed pair of hooks situated on the dorsal surface of the abdomen.

In still and sluggish water there are a number of common water beetles, the most important being the great water or diving beetle, *Dytiscus marginalis*, the great silver water beetle, *Hydrophilus piceus*, and the whirligig beetles.

Dytiscus is one of the most ferocious of all insects, attacking and killing creatures much larger than itself, including small fish. It lays its eggs in spring, the female depositing them in slits which she makes in the stems of water plants. In a short while the larvae hatch, and grow steadily until they are about two inches long. They are just as ferocious as their parents with even greater appetites, because of course their function is to attain full size as soon as possible. Their method of feeding is different from that of their parents, however. The larval jaws are pierced by fine canals which inject a powerful digestive saliva into the victim's tissues, which soon become digested. The fluid products of this external digestion are then sucked up through the same canals into the mouth. The larval mandibles that captured the prey are relaxed only when the victim has been reduced to little more than an empty husk. After about six weeks of intensive feeding the larva is full grown. It now makes its way to the edge of the water, crawls out and buries itself in the soil. Here it pupates, to emerge in a few more weeks as a full-grown adult water beetle.

Hydrophilus is even larger than *Dytiscus,* but it is completely vegetarian in its feeding habits, spending most of its time crawling among the aquatic vegetation on which it feeds. The females lay their eggs in groups of about fifty contained in a covering of silky material. Unlike their parents, the larvae that hatch from these eggs are carnivorous, feeding for choice upon water snails. By the time they reach the last instar they may be as much as three inches in length. Before pupation they leave the water and bury themselves in the soil nearby.

Whirligig beetles are much smaller than the other two kinds, the adults averaging a quarter of an inch in length. They spend their lives near the surface, gyrating round and round with a kind of crazy motion, presumably gathering tiny food particles from the surface film. The eggs are laid in spring, attached to submerged plants in batches of twenty to thirty. Unlike those of other water beetles, the larvae of the whirligig beetles are truly aquatic, being provided with ten pairs of gills which can extract oxygen from the water.

Towards the end of July the full-grown larvae climb out of the water on plant stems and pupate. A few weeks later the adults emerge, and from then until the cold autumn weather sets in whirligigs are particularly common. When it becomes really cold the adults bury themselves in the mud at the bottom of the pond, where they hibernate until the following spring.

Scavengers, animals that dispose of dead animals and plants, are of great importance because they prevent the unpleasant consequences of the accumulation of decaying organic material. Quite a number of beetles are specialists in scavenging: among the best known are the burying or sexton beetles, which specialise in the carcases of dead birds, mice and other small mammals. When they find a dead body they proceed to burrow underneath it and remove the soil on which it is lying, with the result that it gradually sinks into the excavation, until finally it is completely below the surface. The soil that has been excavated is then piled on top of it so that it is completely buried. The female beetles now lay their eggs on the carcass, so that when they hatch the larvae have an abundance of animal food on which they can live until they pupate.

The closely related burrowing beetles are a large group chiefly represented by the dung beetles, which lay their eggs in burrows beneath cow, horse or other manure, plugging the hind end of the burrow with dung on which the larvae will feed when they hatch. The sacred scarab beetles of ancient Egypt belong to this group.

Also belonging to the same group are the chafers, which feed upon plants. The larvae live in the ground, where they feed upon grass roots, and here they also pupate. When you are weeding the garden in early summer you may see an adult pushing its way out of the soil.

Beetles include their full share of insect pests. One of the most notorious of these is the dreaded wireworm, which is the larva of a little beetle known as the click beetle. Click beetles derive their name from the fact that if they land on their backs they are able to right themselves by giving a sudden jerk which produces a sharp clicking noise and sends them up into the air and enables them to turn over. Their yellowish larvae live in the ground and normally feed on grass roots, but if the ground is ploughed and sown with crops they transfer their attentions to the roots of the crop plants. In ploughed grass land they are usually present in enormous numbers. A light infestation representing less than three hundred thousand to the acre will enable almost any crop to be grown without any noticeable loss, but in excess of one million to the acre wireworms pose a serious problem. These click beetle larvae may live in the ground for as long as five years feeding on plant roots before they pupate.

Several groups of beetles specialise in wood-boring larvae. The most notorious of these is the death-watch beetle, *Xestobium rufovillosum*. It is, too, one of the largest of them, although its larvae are only about one-third of an inch in length. They bore into decaying oak, willow, hawthorn and other hardwood trees, and also into the structural timbers of buildings, where their depredations can be very serious.

Equally important economically is the furniture beetle, *Anobium punctatum,* commonly known as the woodworm. Although its larvae are only about one-fifth of an inch in length, they can reduce an article of furniture to little more than powder if allowed to proceed unchecked. The larvae of many other species of beetles live by burrowing into the trunks of growing trees, and these form an important part of the diet of woodpeckers. All these wood-boring larvae feed on wood, being equipped to digest cellulose.

Quite a number of other beetles are economically destructive in their larval stages. *Dermestes* species are extremely damaging to furs and skins in store, as well as to certain stored foodstuffs such as bacon and cheese. A smaller relative called the museum beetle (*Anthrenus musaeorum*) can cause havoc among animal specimens in museums unless adequately guarded against. Another important beetle, which feeds in the larval stage on flour, is *Tenebrio molitor,* whose yellowish-red larvae are the famous mealworms used as food for small reptiles, amphibians and small mammals in most zoos and by all who keep these animals as pets at home.

One family of beetles far exceeds in numbers of species any other family within the animal kingdom. It is the family Curculionidae, the weevils. The world total of species exceeds forty thousand. Weevils can be easily distinguished from all other beetles because the front of the head is formed into a pronounced beak or rostrum which bears the mouth parts. Both in their adult and larval stages weevils are plant feeders, many of them confining their attentions to seeds, making them important economically.

To the forester one of the most detrimental of all insects is the pine weevil, *Hylobius abietis.* The adults attack all conifers, but are particularly partial to Scots pine. They eat into young branches right through to the sap-carrying tissues, and in this way interfere with the passage of sap and so with the growth of the tree. Curiously enough the larvae do not live in the trees, but confine their attentions to felled trunks and decapitated stumps, an accumulation of which can therefore provide breeding grounds for mass attacks on the living trees.

Nut weevils confine their attentions to nuts. The adult female uses her very efficient rostrum to bore a hole right into the centre of the young nut. Having done so she turns round and deposits a single egg in the bottom of the cavity. As the nut grows all external evidence of the original boring becomes obliterated, the entrance hole having healed up. By the time the nut is ripe and falls to the ground, most of its substance will have been eaten by the growing larva that hatched from the egg. This now proceeds to bore a small exit hole in the shell by which it leaves to pupate in the ground. The hazelnut that we crack only to find the withered remains of a kernel is evidence of the activities of a nut weevil.

Closely related to the weevils and equally destructive are the bark beetles, comprising the family Scolytidae. They form complex systems of tunnels between the bark and wood of trees and shrubs. The insects themselves are very small, and can often be more easily identified by the characteristic patterns of their tunnels than by examining the living insects. Bark stripped from a dead elm trunk will often reveal the galleries made by the great elm bark beetle *Scolytus destructor*. The female beetle bores through the bark and excavates a main egg gallery, so called because at intervals along the wall she gnaws little recesses, laying an egg in each. The larvae that hatch from these eggs proceed to gnaw their own tunnels, which all lead away from the main gallery. All the time the larvae are feeding on the sugars and starch contained in the wood. When full-grown each larva excavates a somewhat wider end chamber in which it pupates. The adults that eventually emerge continue to burrow until they break through the surface of the tree.

Although they are mostly quite small insects, the leaf beetles belonging to the family Chrysomelidae are attractive creatures, many of them brightly coloured with a metallic sheen. As their name suggests they feed on the leaves of plants, and many of them do considerable economic damage. The most dreaded of them all is the notorious Colorado potato beetle, *Leptinotarsa decemlineata*, which if unchecked is able to decimate complete potato crops. It was accidentally introduced from America to France in 1922, since when every European country has been constantly on its guard against the introduction of this dreaded pest. Other destructive pests in this family against which the horticulturist is waging perpetual war are the asparagus beetle and the various flea beetles such as the so-called turnip fly.

A few species of *Donacia*, belonging to the same family, have remarkable aquatic larvae. The adult beetles live on waterside vegetation, but lay their eggs on the under-sides of leaves of aquatic plants. As soon as they hatch the larvae migrate to the roots of these plants, and here, although submerged in water, they are able to breathe air. Special hollow spines at the end of the abdomen pierce the tissues of the roots as far as the air spaces. Air from these passes through the spines and into the tracheal system of the insects.

In contrast to so many of the beetles we have so far met, our last group, the ladybirds (family Coccinellidae) are among man's most useful allies in his fight against insect pests. Both in their larval and adult stages ladybirds feed exclusively on aphids. The females lay their eggs beneath the leaves of any plant on which there is a flourishing colony of aphids, so that from the moment they hatch out the larvae have a plentiful supply of food available to them. To deal with such soft prey they do not need very powerful mouth parts. When they are ready to pupate the larvae merely crawl to the under-surface of a leaf. The pupae look rather like dried up larvae, because they remain covered with the shrivelled larval skins. In late autumn, as aphids become scarce, the surviving adult ladybirds seek out deep cracks in bark, and here they go into hibernation until aphids become active again the following spring.

CHAPTER SIXTEEN

ARACHNIDS, CHILOPODS AND DIPLOPODS

The subphylum Arachnida contains a great variety of types. The most numerous and familiar are the spiders, which can be found in virtually every part of the world and in every habitat. Ticks, mites and scorpions are also pretty well known. The subphylum Chilopoda comprises the centipedes, and the subphylum Diplopoda the millipedes. Until recently these two kinds of arthropods were classed together in one subphylum, the Myriapoda.

Mating in spiders is a fascinating process, but in order to understand it, it is necessary to have some knowledge of the relevant anatomy of the male and female spider. In the female the two oviducts merge just before opening in a single aperture towards the front end of the under-surface of the abdomen. Just behind this opening are two other apertures leading into a pair of sperm storage organs or spermathecae.

In the male spider two openings from the paired testes occur in a similar position on the under-surface of the abdomen. The mating process is quite unlike that in any other animals except the millipedes, which have a somewhat similar method. Behind the pair of chelicerae or jaws all spiders have a pair of pedipalps, which are essentially sense organs. But in the male the terminal segment of each pedipalp has become modified to play a most important part in the whole mating procedure.

Soon after undergoing its last moult the now fully adult male proceeds to spin a strip or triangle of silk a few millimetres long. On to this he deposits a drop of seminal fluid by

Tarantula courtship

Tarantulas mating

tapping his genital openings on it. He then turns round and applies the tips of his palpal organs to it, and sucks it up in much the same way as a fountain pen filler sucks up ink. Thus charged he goes in search of a female, and after the necessary preliminaries he inserts his charged palps into the openings of her spermathecae, into which the seminal fluid is transferred.

The preliminaries, however, are extremely important. Before he approaches too closely he must ensure that the female has recognised him as a potential mate, or she will almost certainly regard him as prey and his amorous advances will come to an abrupt and tragic end.

The courtship antics by which he achieves this essential recognition vary according to the general habits of the group to which he belongs. They are in fact closely related to the methods by which members of his group normally detect their prey. Jumping spiders and wolf spiders rely entirely upon sight to recognise and pursue their victims, and consequently the males of these species make visible recognition signs to the females of their choice while standing a discreet distance away from them, only approaching more closely when they receive an answering gesture of welcome. In nocturnal species and species with very poor eyesight the male taps the female with his front legs, and she is apparently able to interpret and respond to the message. Males of the web-spinning species communicate with the female in her web by a kind of morse signalling by plucking one of the main radii of the web.

Once his signal has been acknowledged and he knows that he can approach the female safely he proceeds to transfer the seminal fluid from his palpal organs to her spermathecae. In some species he does this by crawling under her body from in front, while in others he climbs on to her back, leaning over to one side to insert one palp into her spermatheca on that side, and then over the other side to insert the other palp into her other spermatheca. Having completed the mating process the male usually retreats fairly smartly, because there is always a danger that if he lingers the female may attack and kill him. Cannibalism is a fairly well developed instinct in spiders.

Having successfully concluded one mating the male now spins another silk strip and recharges his palps before going off in search of another mate. How many times he repeats this procedure is not known, but it probably varies from species to species.

Once safely inside the female's spermathecae the spermatozoa can remain alive almost indefinitely until they are required, even from one summer right through the winter until the following spring. The females of most species lay more than one batch of eggs, and some species may produce a considerable number of batches. The first batch may be laid very soon after mating, or there may be a considerable delay. However many batches are laid, the spermatozoa obtained at a single mating are sufficient to fertilise them all.

Before she lays a batch of eggs the female spider first spins a saucer-shaped mat of silk, somewhat similar to that produced by the male before he discharges his seminal fluid. As the eggs pass out of her oviducts sufficient seminal fluid is discharged from the spermathecae to fertilise them. The fertilised eggs are then covered with layers of silk, thus converting the saucer into a cocoon. In some species the mother mounts guard over the cocoon until the eggs have hatched. Many of the web-spinners construct elaborate cocoons that are suspended either from the web or from a nearby twig. Female wolf spiders attach their globular cocoons to their spinnerets and carry them about with them wherever they go, and the young spiders when they hatch climb on to their mother's back, where they remain for several days before finally leaving her to lead independent lives.

The size of spiders' eggs varies much less than the size of the spiders themselves, so that the eggs of small spiders are relatively bigger than those of larger species. As a result small spiders lay fewer eggs in each batch than large ones. Very small species may only produce two or three eggs in a batch, whereas large species may produce between two and three thousand in a single cocoon.

Newly hatched spiders have no hair or spines, and they are

unable to feed. In a few days they moult for the first time acquiring hair and spines and the ability to feed themselves and to produce silk. Only now do they usually bite their way out of the cocoon, although if the weather is bad or the season not right they may remain within the cocoon for a matter of weeks or even months.

A few kinds of female spiders show active parental care and provide their newly hatched babies with food. The female of one species, *Theridion sisyphium,* hangs upside down in her web and regurgitates liquid food which the young spiders take from her mouth.

But the majority of spider mothers cease to concern themselves with their offspring after they have hatched and undergone their first moult. By this time they have used up all the food contained in the egg yolk and are hungry. Dispersal now becomes an urgent necessity, because if the young spiders remain together their inherited cannibalism will take control and they will begin to eat one another. And so they begin to scatter. In some species this merely involves walking from the area of concentration, but many others disperse by the spectacular method of 'ballooning'.

The little spider climbs as high as it can on an exposed grass stem or other plant. It then turns to face whatever breeze there may be, raises its abdomen in the air and secretes a tiny globule of silk from its spinnerets. This is drawn out into an incredibly thin thread by the breeze, and when it has reached perhaps a yard or so in length it has sufficient buoyancy to lift the spider from its anchorage and carry it into the air. It is possible for small spiderlings to be carried for several miles on these gossamer parachutes, and at least on some occasions for much greater distances, before finally coming to earth. Darwin, in his famous account of the voyage of the *Beagle,* records several thousand coming aboard on 1 November 1832 when the vessel was sixty miles from the nearest land off the mouth of the River Plate.

An interesting theory seeks to explain the origin of the word gossamer, originally applied to these threads but subsequently extended to describe anything particularly light and thin. According to this the term was originally goose-summer, and alluded to the fact that these floating threads are most abundant at the Michaelmas season, the traditional time for geese to be eaten—they would be purchased at the many goose fairs which used to be common at this time of the year in earlier times.

Scorpions are essentially creatures of warmer climates, many of them being well adapted to living in desert conditions. Mating is preceded by elaborate courtship dances, at the end of which the males deposit spermatophores on the ground and the females move over them and take them into their genital openings.

Fertilisation occurs internally, and the fertilised eggs are retained within the female oviduct until they have hatched. In some species they are able to subsist on the yolk originally laid down in the egg, but in others the developing egg comes into intimate contact with the wall of the oviduct, from which it is able to absorb food materials, the whole arrangement constituting a kind of primitive placenta.

When eventually the young are born they climb on to their mothers' backs, where they remain at least until they have moulted for the first time, and often for a further week or two. The mothers make no attempt to feed them, however, but they can exist over this period by using up the food materials provided in the yolk of the egg. The total number of moults occurring before the scorpions become adult is not known, but probably in most species it is something like eight.

The third important group of Arachnida contains the mites and ticks. Their bodies show no differentiation into cephalothorax and abdomen, the whole body being ovoid with four pairs of thoracic limbs towards the front. One of the best known members of the group is the sheep tick, *Ixodes ricinus*, which because of its shape is sometimes also known as the castor bean tick. The head end with its mouth parts is buried in the host's flesh, from which it obtains its supplies of blood.

The sheep tick has a complicated life history typical of the members of the group. Mature females lay their eggs on the ground, and from them hatch larvae with six legs. These climb up to the top of grass stems, where they lie in wait for a passing sheep. If a sheep brushes against it the larva will attach itself to one of the hairs, passing down it until it reaches the skin, into which it immediately plunges its mouth parts. Retaining a firm hold, it will suck in its host's blood until it is gorged, when it will release its hold and drop back to the ground.

As its meal is absorbed, so it moults and becomes a nymph with eight legs. Once more it climbs up a grass stem, and again attaches itself to a passing sheep, from which it takes a second large meal, and again detaches itself and drops to the ground to digest the meal and moult. This time a mature tick emerges to repeat the same method of attaching itself to a host. Mating between the males and females takes place on the host's body, followed by the female detaching itself and dropping to the ground, this time to lay its eggs. The chances of the tick's being able at any stage to attach itself to a passing sheep are very slender; the majority of the waiting larvae, nymphs or mature adults perish on their grass stems, and only a tiny minority is successful in finding the host they are seeking. It is to compensate for this that enormous numbers of eggs are produced.

Water mites are among the most common mites. They are quite common in almost any stretch of fresh water. Most of them are active creatures, swimming quite powerfully through the water or climbing among the water plants in pursuit of their prey, which consists of almost any kind of tiny animal life. Water mite eggs hatch to a larval stage in which there are only three pairs of legs. All these larvae, so far as is known, are parasites of some other aquatic creature. As soon as possible after hatching a larva will seek out its particular host, into which it buries its head so as to be able to feed on its host's blood or body fluid. Its legs, being of no further use to it, degenerate and drop off. Meanwhile, within the larval body the next stage is developing, and this eventually emerges as a fully formed nymph with four pairs of legs. Its appearance is similar to that of an adult, and it leads a free-swimming existence actively pursuing its prey. After a time its activity ceases and is replaced by an inactive resting phase, during which an adult mite develops beneath the nymphal skin, from which it emerges as soon as it is fully formed.

A large group of soft-bodied mites live on various foodstuffs. There are for example the sugar mites and the flour mites, but perhaps the best known members of this group are the cheese mites, which in fact do live on other foods as well as cheese. Their method of reproduction is by viviparity, the females producing young instead of eggs, and this means that their rate of reproduction is fast. Because of this a colony of cheese mites can consume a piece of cheese in quite a short time.

Closely related to the food mites are the irritating itch mites, which bury themselves in our skin and cause intense irritation. The females excavate tunnels just beneath the skin. These may be an inch or two in length, and are packed with fertilised eggs. When these hatch the larvae can be transferred by contact either to other people or to cats, dogs, horses and cattle.

Superficially, millipedes and centipedes certainly appear to be similar. They have long, thin, flexible bodies and many pairs of legs. But the differences between them are much more fundamental. For example the abdominal segments of the millipede each carry two pairs of legs, whereas those of the centipede have only a single pair, but they are much better developed. All centipedes are carnivorous, and the appendages of the first abdominal segment are modified to form a pair of poison claws with which their prey is killed. Millipedes possess no such weapons, and are mainly vegetarian in their feeding habits.

The reproductive openings in both sexes of the millipedes are either on the third segment or actually on the second pair of legs. But, as in the spiders, mating does not involve a direct confrontation between the male and female reproductive organs. In the male millipede the appendages of the seventh segment are modified to form organs called gonopods, which function like the terminal segments of the pedipalps in spiders. Before mating, the male millipede charges these gonopods with seminal fluid by flexing his body so that the gonopods come into contact with the openings of his testes on the third segment. Actual mating consists of the application of the charged male gonopods to the openings of the female oviducts and the transfer of seminal fluid to them.

The fertilised eggs are laid in the soil. Sometimes they are coated with soil particles and excrement and deposited in crevices in the soil; but in other species they are laid in specially constructed nests. These are made of soil particles moistened with saliva, and are in the form of hollow spheres. They are lined with excrement and are smooth inside but rough and irregular on the outside.

Those that lay their eggs singly in the soil generally lie on their backs and pass their eggs backwards along their numerous appendages until they are over the anus, when the rectum is everted to give them a covering of very fluid excrement. As this dries it forms a tiny hollow chamber in which the egg lies free. Sometimes two or three eggs may be enclosed in a single chamber, but in this case they always occupy separate compartments within the chamber.

The larvae undergo a series of moults before they become adults, and at each moult a few more segments and legs are added between the front segments and the anal segment.

Although centipedes are very common animals, much remains to be learned about their methods of reproduction. In contrast to the millipedes, the first larval stage in the majority of species emerges from the egg with the full adult number of segments.

Mating in centipedes involves the formation of spermatophores by the males, which, after they have been shed, are picked up by the females. In some species the spermatophore is deposited on a tiny web of silk.

Parental care is very common. In many species the eggs are merely deposited in the soil in masses containing between fifteen and thirty-five eggs. While they are developing the female often licks them, and in doing so is believed to be covering them with a fungicidal secretion. Left unattended, mould spores are very likely to germinate on them. In the common and widespread Scolopendromorpha the mother often hollows out a brood chamber in rotting wood, and here she curls her body around the developing eggs, lying on her side so that the eggs are cradled between her numerous legs.

One of the difficulties about investigating the breeding activities of centipedes is that as soon as she is disturbed the brooding female will either eat her eggs or abandon them, when they soon become attacked by fungi.

CHAPTER SEVENTEEN

ECHINODERMS

The members of the phylum Echinodermata are interesting in themselves, but from the evolutionary point of view they are extremely important as being the probable ancestors of the Chordata. They are also particularly interesting because of their very unusual structure. Except for the primitive coelenterates, the sea anemones and the jelly-fish, the members of all other phyla are built on the bilateral plan, in which an animal has a front and a hind end, with the exception of the echinoderms, which share with the coelenterates the distinction of being the only two groups of animals built upon the radial plan.

The phylum contains five classes which are readily distinguished in external appearance and which show significant variations in their methods of reproduction, especially in the kinds of larvae they produce.

The best known are the starfish, forming the class Asteroidea. The starfish body consists of a small central disc from which, typically, five tapering arms grow out, though in a few species there may be more than this basic number. These arms are much more than appendages for locomotion: there is insufficient room in the central disc for all the necessary internal organs, so the arms provide room for some of them, housing the reproductive system and part of the digestive system.

If you pick up a starfish you will notice that it has a soft, almost flabby body, but that its skin is rough, being covered with numerous tiny outgrowths. These are short blunt spines which are based on tiny calcareous plates embedded in the skin and forming a loose skeleton. It is the possession of these skin spines that gives the phylum its name—echinoderm means 'spiny-skinned'.

Perhaps the most unusual feature of the echinoderms, apart from their radial symmetry, is their method of locomotion. If you watch a starfish moving about on the bottom of a rock pool it will seem to glide slowly along without moving its body. If you turn it over you will discover the secret of its locomotion. A groove runs right along the under-side of each arm from the central disc to the tip. On either side of this groove there is a row of small tube feet each ending in a flat disc which can be used as a sucker. These tube feet are hollow and open internally into a canal which is filled with water and runs along the whole length of the arm. The movements of the feet are controlled by pumping water into and out of them from the canal.

Living on the sea bed the starfish is liable to get its arms caught under stones rolled by the waves. Like the crab it can free itself by amputating the trapped arms and growing new ones. It has in fact great powers of regeneration. Sometimes only one arm and a portion of the central disc can be torn away, but even from this small remnant a complete starfish can be regenerated.

Similar to the starfish but easily distinguished from them are the brittle stars, forming the class Ophiuroidea. In contrast to the true starfish their central disc is clearly marked off from the arms, which are relatively thinner, much longer and more flexible than those of the starfish. Lateral rows of spines grow out from the arms, which have no groove on the under-surface. The tube feet are small and pointed, and without suckers. Brittle stars walk on the sides of their tube feet, which secrete a sticky mucous which gives them some degree of adherence. Unlike true starfish they can swim through the water by gentle undulations of their flexible arms.

The third echinoderm class, the Echinoidea, comprises the sea-urchins. At first sight there may seem little resemblance between the flat starfish with its prominent arms and the extremely prickly ball that is the sea-urchin. Yet the structure of the sea-urchin may be regarded as the result of a drastic experiment with the typical starfish structure. It can best be understood if we imagine the five arms of the starfish turned upward and inward so that the tips meet above its body and the five spaces in between filled in, the whole thus forming a hollow ball.

The long spines of the sea-urchin are comparable with the much shorter spines of the starfish. A careful examination of this globe will show that, starting from near the mouth, which opens on the lower side, five double rows of tube feet run nearly up to the top of the globe. The sea-urchin uses these in the same way as the starfish uses its feet to glide along the sea bed. If it should fall it will not matter how, because some of the tube feet will be near to the ground, and it will therefore be able to move and right itself.

Less well known than the three previous classes are the members of the fourth, the Holothurioidea or sea-cucumbers. Although their bodies are elongated and worm-like in general appearance, they are built on the same fundamental plan as those of the sea-urchins and the starfish. The mouth surface is carried at one end of the cylindrical body instead of underneath, so that the five rows of tube feet run back along the body. The tough skin is similar to that of the starfish, and is likewise strengthened with isolated plates. The mouth is surrounded by a ring of ten branched tentacles.

Sea-cucumbers are suspension feeders with an unusual method of procuring their food. The tentacles are covered with a sticky fluid which traps the organisms suspended in the water. Instead of these then being carried to the mouth by means of cilia, as is usual with suspension feeders, the tentacles are themselves periodically curved round into the mouth and withdrawn through another special Y-shaped tentacle, which scrapes off the food much as a child might lick jam off its fingers.

Among the seaweeds covering the rocks on warmer coasts may be found an interesting creature called a feather star, which differs from starfish and brittle stars in several important respects. It has ten thin jointed arms from which numerous tiny branches grow out, giving them the appearance of feathers. There are no tube feet, and the feather star swims about in the water using its arms as oars. These always work in pairs, because each pair represents one arm which has divided into two parts at its base.

The mouth surface is carried uppermost, in contrast to the starfish and brittle stars, and rows of cilia along the upper surface of the arms cause a current of water to flow inwards towards the mouth. From this water it extracts its food, consisting of suspended plankton organisms.

The feather star is of special interest because it is a survivor from a very remote past, when vast areas of the deep sea bed were covered with creatures like feather stars growing at the tops of long stalks permanently anchored to the bottom. Many of the stalks grew to twenty feet or more in height. The mouth surface was carried upwards, and they obtained their food from water currents caused by beating cilia along their arms just as the feather stars do today. They are known as sea-lilies or stone-lilies because they looked like enormous grotesque flowers. In time some of the sea-lilies acquired the habit of breaking away from their stalks when they were full grown to lead a free-swimming existence for the rest of their lives. The modern feather stars are descended from these types, and as we shall see their development betrays their ancestry. Sea-lilies and feather stars constitute the fifth echinoderm class, the Crinoidea.

Although their period of greatest abundance occurred in much earlier times, the fixed sea-lilies are far from approaching extinction. Since, however, they flourish mainly in very deep waters they are seldom encountered. In this connection it is interesting to recall that the famous *Challenger* expedition of the 1870s brought up from depths of between two and three thousand fathoms specimens of sea-lilies which had previously been known only from fossils and which were believed to be extinct.

Even in their development the echinoderms are unusual. The fertilised eggs of all five classes give rise to bilaterally symmetrical larvae which have a front and a hind end, each class having its characteristic larval form. The larva produced by a starfish is known as a bipinnaria, a rather grotesque larva having a number of 'arms' jutting out from the main body, the whole being provided with two separate bands of cilia encircling the body. These bands, since they follow the outline of the larval arms, are much convoluted. There is a mouth at the front end and an anus at the rear. For a time these larvae live in the plankton, where they feed and grow, swimming by means of their bands of cilia.

Eventually comes the time when they undergo a metamorphosis which converts the bilaterally symmetrical larva into a radially symmetrical adult, and this is achieved in a most unusual manner. Towards the right side of the larval body the five true adult arms develop, the outer surface being the upper (dorsal) surface of the adult, the ventral surface with its mouth developing on the left side of the larva. As soon as these adult rudiments are sufficiently developed the larval arms, together with the remainder of the larval body, are absorbed into the new adult body. In a few species these larval structures are actually shed. In either case all signs of the original bilateral symmetry are obliterated.

The auricularia larva produced by the fertilised eggs of the sea-cucumbers is not unlike the starfish bipinnaria larva in general appearance, but to begin with it has only a single band of cilia. As development proceeds the larval arms, instead of elongating as they do in the bipinnaria, become reduced, and the single convoluted ciliary band becomes broken up into a number of separate simple bands encircling the larva at different points along its length. This second stage larva is sometimes called a doliolaria. In some species, instead of swimming smoothly by means of its ciliated bands like bipinnaria and auricularia larvae, the doliolaria has been seen spinning like a top. Development of the adult occurs in the doliolaria in the same way as it does in the bipinnaria larva.

Sea-urchins and brittle stars produce echinopluteus and ophiopluteus larvae respectively, and these differ from the bipinnaria and auricularia larvae in having greatly extended slender larval arms which are supported by an internal skeleton consisting of rigid rods. The course of metamorphosis, however, is the same.

The larval life of most starfish, sea-cucumbers, brittle stars and sea-urchins is usually prolonged, and may well last for several months. By contrast, the complicated metamorphosis is often completed in twenty-four hours or less. In some brittle stars and sea-urchins it is said that the whole process may take less than an hour. In the vast majority of species metamorphosis occurs while the larvae are still drifting in the plankton, but some starfish larvae sink to the bottom and become temporarily attached by means of a sucker while the minute adult body develops. This then tears itself away from the attached larval structures.

Fertilised feather star and sea-lily eggs develop first to form a doliolaria larva, which has a tuft of very long cilia at the anterior end and four bands of normal short cilia surrounding the body at different levels. In contrast to the larvae of the other four classes this doliorlaria exists in the plankton for only a few days before settling on the sea bed, where it fixes itself by its anterior end to rocks, stones or weeds and undergoes a rapid change. The body elongates considerably to form a stalk, while at the free end, which is the original posterior end, a structure called the crown or calyx develops. This has the typical structure of a feather star with five branched arms and a mouth in the centre of the upper surface.

This stage is known as a pentacrinoid larva. In the sea-lilies the larva merely increases in size, the stalk becoming longer and stronger as the calyx grows to full size, but in the feather stars the calyx soon breaks away from its stalk to swim free in the water for the rest of its life, the stalk being left behind to die.

In common with other marine animals that shed their reproductive products into the sea and rely upon random

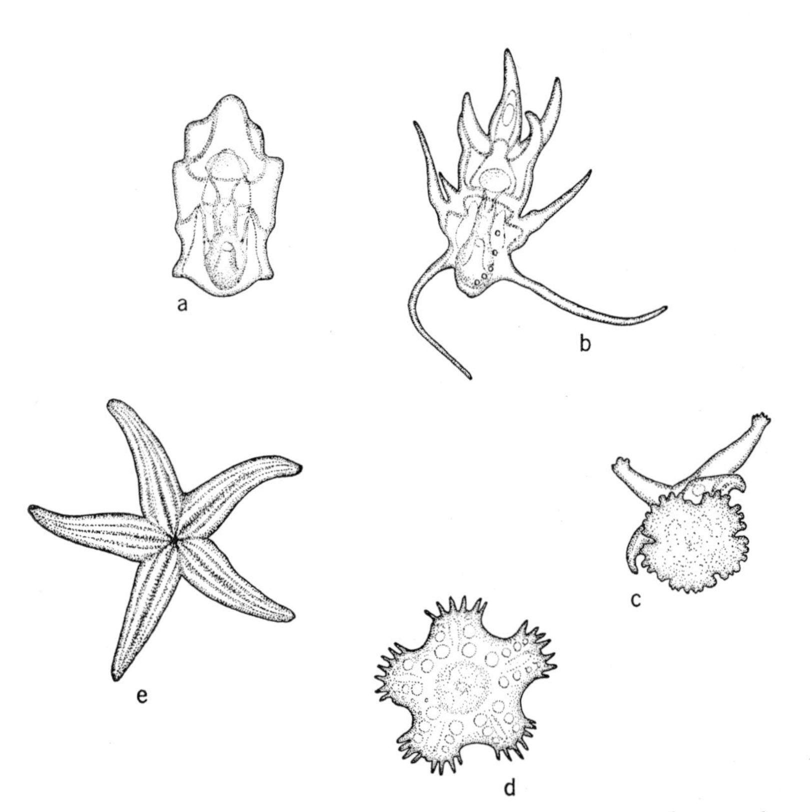

Development of the common starfish, *Asterias* (not to scale to each other but drawn to show relative shapes): *a* bipinnaria larva; *b* brachiolaria larva; *c* late stage in development of brachiolaria larva into young starfish; *d* young starfish; *e* full-grown *Asterias*

fertilisation of their eggs, the echinoderms produce enormous quantities of these products. The female common red starfish, *Asterias rubens*, is believed to be capable of passing out as many as two and a half million eggs into the water in a couple of hours when it is spawning. The sexes are separate in most echinoderms, and to ensure that the males shed their spermatozoa at the same time as the eggs are appearing it seems that the presence of eggs in the water stimulates the males to spawn.

A few species in each class have abandoned the broadcast

method of reproduction, and instead of producing enormous numbers of tiny eggs they produce relatively few. Each of these, however, is a large egg supplied with quantities of yolk containing sufficient food materials to carry the embryo right through its larval development and metamorphosis, so that it has no need to feed. The majority of the larvae produced from these large eggs enjoy the benefits of some form of parental protection during their development. Most of the echinoderms that indulge in this parental protection live in polar regions, particularly the Antarctic. Echinoderms that retain the broadcast method of reproduction usually have a restricted breeding season, generally confined to the spring, but those species that have developed some kind of parental care usually have a much extended breeding season. Indeed, they will often be found breeding all round the year.

In the crinoids, which protect their broods, each ovary has beside it a brood pouch or marsupium in which the fertilised eggs develop. Only when they are ready to change to pentacrinoids do they leave these pouches and pass out into the sea.

Starfish show two different methods of protecting their developing larvae. Some long-armed species hunch their bodies over the developing eggs so that they come to lie under a kind of tent beneath the bases of the arms and the mouth. While they are thus engaged in protecting their developing young the adult starfish are unable to feed. By contrast, certain Antarctic species belonging to the family Pterasteridae form a kind of brood pouch on their upper surfaces, the skin being pushed down to form a hollow which is then roofed over by folds of the surrounding skin. Correlated with this habit the reproductive ducts of the Pterasteridae open on the dorsal surface and not ventrally as is usual in starfish.

A few holothurians incubate their special large eggs in the body cavity, though how these eggs are initially fertilised is not known. Some other species protect their developing eggs in skin pockets in their upper surface, in a manner similar to that adopted by the Surinam toad, *Pipa americana*. In some species each pocket contains only one developing embryo, whereas in others there may be several. How the eggs become embedded in the skin is not known, but presumably they are aided by the tentacles or the tube feet of the female.

In the large Antarctic brittle star *Ophionotus hexactis* a small number of larvae develop in the female's central disc. This may be no more than two inches in diameter, yet the developing young within it may attain a diameter of three eighths of an inch before passing out through the mouth to lead a free and independent life.

Certain species of sea-urchins also protect their developing young. In some the young nestle among the spines surrounding the mouth on the lower surface, while in others they occupy a similar position at the apex of the upper surface.

BIBLIOGRAPHY

It is not possible to give a list of books dealing specifically with animal reproduction, but the following selection is a guide to further reading.

Bellairs, A. *The Life of Reptiles.* 2 vols. Weidenfeld and Nicolson, 1970

Buchsbaum, R. *Animals Without Backbones.* 2 vols. Pelican Books, 1951

Clark, A. M. *Starfishes and their Relations.* British Museum, 1968

Cloudsley-Thompson, J. L. *Spiders, Scorpions, Centipedes and Mites.* Pergamon Press, 1968

Marshall, N. B. *The Life of Fishes.* Weidenfeld and Nicolson, 1965

Matthews, L. H. *The Life of Mammals.* 2 vols. Weidenfeld and Nicolson, 1969 and 1971

Morton, J. E. *Molluscs.* Hutchinson, 1958

Oldroyd, H. *Elements of Entomology.* Weidenfeld and Nicolson, 1968

Richards, O. W. *The Social Insects.* Macdonald, 1953

Romer, A. S. *Man and the Vertebrates.* 2 vols. Pelican Books, 1954

Smith, E., etc. *The Invertebrate Panorama.* Weidenfeld and Nicolson, 1971

Street, P. *The Crab and its Relatives.* Faber and Faber, 1966

Vesey-Fitzgerald, B. *The Worlds of Ants, Bees and Wasps.* Pelham Books, 1969

Young, J. Z. *The Life of Vertebrates.* Oxford University Press, 1950

In addition many of the volumes in Collins New Naturalist Series will give additional background information.

INDEX

Note: page numbers in italics indicate illustrations